AN ACTIVE HAND

Fundamentals of Restoration Forestry

Peter P. Bundy

The Society of American Foresters
10100 Laureate Way
Bethesda, MD 20814, United States
www.eforester.org

Disclaimer: This book and contents included within do not constitute professional advice and should not be taken as such.

Printed in the United States of America on paper certified by the Sustainable Forestry Initiative. The paper used in this printing meets the requirements of ANSI/NISO Z39.48-1992 (Permanence of Paper).

ISBN: 978-0-939970-64-3

An Active Hand: Fundamentals of Restoration Forestry

CONTENTS

CONTRIBUTORS

The Society of American Foresters thanks the below-named peer reviewers for their thoughtful and critical reviews of the manuscript.

Lauren Grand
Assistant Professor of Practice, Forestry, and Natural Resources Extension
Oregon State University
Eugene, OR

Klaus Puettmann, PhD
Oregon State University
Corvallis, OR

Marcella Windmuller-Campione, PhD
Assistant Professor of Silviculture
University of Minnesota
Saint Paul, MN

INTRODUCTION

Welcome to a disturbed and degraded landscape, one that is at once wounded and yet full of potential. It is a landscape where patience and perseverance are necessary components on the path to restoration. This book is for those who understand that as human beings, we are a critical component of the natural world. We do not stand apart. Nature is not seen as resting under our dominance, as she had been in the past. But nor is she, as some would argue today, better than we are. We are partners and need to work together with nature to make forest restoration happen.

An Active Hand posits that the human role, while historically disruptive, is ripe to become that of active healing and that mere preservation is not sufficient. This is not, however, a textbook on forest restoration, for those have already been written. It is instead a guidebook that is based upon 30 years of field experience; specifically work in research, fieldwork in collaboration with public sector managers; and sustained assistance to family woodland owners in diverse woodlots across the Midwest.

It is necessary, of course, to analyze forest degradation, historic conditions, and the many technical considerations that come with implementation of forest restoration. But this book has much more to offer in that it is derived from a series of real-life stories of successes and failures in the field. These provide a personal context, a *culture*, if you will, to the science and art of restoration.

The text begins with a look at the conditions of our degraded forests today. From there it undertakes to examine the history of ecological restoration in America, and the challenges encountered during restoration efforts of the past. It takes a close look at the causes of degradation in order to understand the history and trajectory of abuse and neglect. These abuses are often man-made, brought about by agricultural practices, severe harvesting, or strip-mining. However, they also derive from natural forces outside of human control, such as windstorms, wildfires, and drought. A single such windstorm, for example, flattened a quarter section of the author's own land. From this sobering experience the author learned that there is a need for a new set of tools to care for the woods, one seldom taught in forestry schools in the past. It is to address this need that he has written this book.

* * *

An Active Hand relies upon three legs. The first is ecological: that mosaic of interconnections between the soils, the plants, and the animals. With ecological knowledge as well as an understanding of prior conditions, we may begin to reconstruct restoration goals.

The second leg is economic. Here, we often falter, for many restoration projects carry a heavy price tag. How do we overcome these economic barriers? What methods may produce income during restoration projects? And what are the long-term economic benefits of restored landscapes, whether on public or private land?

The third leg is cultural and is one that many professional managers neglect. Here, neighbors and colleagues interact and create community. This cultural connection ties humans to the land and gives motivation to difficult restoration decisions. Where community is absent, restoration often falters.

This book recognizes that new and unexpected challenges will inevitably arise. These may come from climate change, from exotic invasive species, or from a host of political issues. They may even originate from our friends. How do we prepare for these challenges? How do we persuade and persevere when the odds are against us?

An Active Hand is the result of 30 years of scientific, silvicultural, and field experience. The narrative, although regional in scope, carries lessons for a wider audience. The argument does not stop at state or national boundaries. It is time for those of us who care to step up to the plate, for the goals of forest restoration are worthy and the process is rewarding.

CHAPTER I

OUR FORESTS TODAY

Beyond a Painful Past

Restoration begins with pain. The pain of loss. The pain of disturbance. The pain of mortality. Across North America, forestlands have been degraded from a painful past. Degraded from early logging practices and subsequent forest fires. Degraded from heavy pasturing and farming. Degraded from recent fires, windstorms, and exotic bugs from distant shores. These abuses have occurred with surprising regularity since Europeans found the "virgin" forests of the Americas and began systematic settlement and exploitation more than 500 years ago.

Human disturbance and degradation are merely the most recent events of a much longer geological pattern on the landscape. If we turn back the clock a few more millennia, what was the status of temperate forests in North America? Much of the northern landscape was altogether devoid of trees. Only 12,000 years ago—the blink of an eye from a geological perspective—the massive ice sheets of the Wisconsin Ice Age crushed the groundcover and left almost half the continent under the residue of glacial till and outwash sand. Imagine, if you will, an expanse of raw tundra thousands of miles wide and the windswept barrens of lichens, rock, water, and sand. Only when the ice retreated north did the slow process of natural regeneration commence. This is the landscape that today I call home, the northern forest (Fig. 1.1).

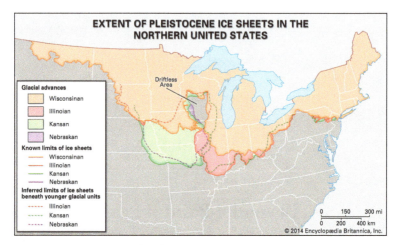

Figure 1.1: Map showing farthest extent of Wisconsin Ice Age.
Reprinted with permission from Encyclopaedia Britannica,
© 2014 by Encyclopaedia Britannica, Inc.

* * *

Homo sapiens and modern degradations are merely the latest wave of a painful past, the latest wave of disturbance ecology in action. Bulldozers mimic miniature glaciers, transforming woodlands to gravel and carving the landscape into housing developments, highways, and shopping malls. White pine (*Pinus strobus*), hickory (*Carya* spp.), and red oak (*Quercus rubra*) have seen this before. Surely this will not be the last time the temperate forest is threatened.

Three hundred fifty years ago, early European settlers in New England and Virginia began systematic destruction of their woodlands. In rocky Massachusetts, stone walls were erected and trees were cut for pasture or for planting corn. In the more fertile southern states, forests were felled and planted with tobacco or cotton. Millions of acres of wild woodland were converted for agriculture and grazing by a young country hungry for food. Degradation and disturbance went hand in hand with human actions.

Even further back on the geological clock, nature was the prime driver of disturbance, as water levels receded from submerged Florida and the Gulf shores. Ice-age gravel on the Canadian Shield became home to lichen, moss, and grass. Grass provided a duff layer to burn. Fire provided carbon, nitrogen, and potassium nutrients to the sterile soil. The grasses flourished and behind them followed moisture. Migration increased, first with insects, then birds, and finally mammals. Humans were the last major arrival, first from Asian, then European shores.

Because the human population has increased dramatically in the past several hundred years, we are in a position to ask, "Do we wish to repeat the painful stories from the past: the demise of Persia, of Rome, and of Spain, three civilizations that exploited their forests and later suffered the consequences of their demise? (*See* Perlin 1989.) Or, will we choose to nurture our forests back to health?"

The choice is ours.

* * *

Restoration forestry is the human art and science of returning forestlands to healthy conditions. Restoration assumes that the landscape has been degraded through past events. These events may have originated from human hands. Degradation may also have originated from other causes, such as drought, windstorms, floods, and severe wildfires. Whatever the cause, biological conditions in the forest become disrupted, and the forest's resilience is reduced, its future cast in jeopardy.

Restoration forestry does not assume that we return the forest to precise presettlement conditions because conditions have changed (Boedhihartono and Sayer 2012). The restoration of a historic home does not exclude modern conveniences of plumbing, lighting, and central heat, nor does it ignore modern building codes, new types of brick and concrete, or triple-paned windows. In the same fashion, restoration of a forested site needs to recognize a climate that is warmer than the recent past, the arrival of new exotic species, and options for amenities such as benches and trails. A return to a "presettlement" past is not preferable or even possible because the ecological landscape has changed.

Restoration forestry does, however, restore historic functions. For example, it includes large-diameter trees as an important component of the forested landscape. It also encourages a variety of active approaches to management. The species mix and desired future condition of restoration prescriptions need to be designed to recognize that historic conditions varied across the landscape. In a nutshell, restoration forestry offers land managers and landowners an opportunity to rehabilitate and restore the full vibrancy and vitality to our forests. Resilience will increase and forests will continue to offer fresh oxygen to breathe, new fiber to use, and clean water to consume, in addition to providing recreational and aesthetic benefits.

This restoration effort will require more than forest science to succeed; restoration is also an artistic endeavor that combines the knowledge of science with the creativity of the human mind. In this manner, restoration represents a Hegelian synthesis of past actions and reactions by honoring the past and innovating for the future.

How Disturbance Drives the System

Restoration begins with an understanding of the natural world, and ecology is the first stepping stone in this process. Just as Plato envisioned a perfect philosopher's state 2,500 years ago, early ecologists designed models that fit their idea of a perfect natural world. Clements, Daubenmire, and Gleason strongly believed that forested ecosystems gravitated toward steady state or "climax" forests. These ideal ecosystems operated, in their eyes, as the pathways for plant species to find their niches without disturbance in a natural world. To make these idealized systems work, fire was removed, windstorms were ignored and humans were not considered by the early ecologists.

But nature does not operate as an abstract model created by philosophers and academics. It operates on its own set of complex variables, variables that the specialists of today, whether ecologists, hydrologists, land managers, or entomologists, are only beginning to understand.

Disturbance is at the center of this system. In fact, disturbance by natural forces drives the system from top to bottom. It is the rare exception, not the rule, that permits the existence of an undisturbed or climax forest, such as the old-growth redwoods (*Sequoia seprevirens*) in California or hemlock (*Tsuga Canadensis*) in northern Michigan.

In most parts of the world, one or more of seven catalytic disturbances shape forest systems: volcanic eruption, fire, wind, drought, insects, nonhuman vertebrates, or humans. All of these variables and disruptions may bring degradation to forested landscapes, both in the short and long terms. Often they act in combination.

If we examine the history of the mixed conifer-hardwood forests of North America, for example, fire was the most common form of disturbance. Because fire leaves charcoal traces of its passing, it has been studied intensively (Egan and Howell 2001). Fire scars are common on old trees, hence early fires may be tracked and chronologically dated. In addition, charcoal, the residue of fire, remains in the topsoil, usually in thin layers. This provides another tool for paleoecologists to develop a fire chronology. These tools, together with anecdotal evidence from European explorers and surveyors, offer a record of the role of fire in postglacial, presettlement North American natural history.

Fire historians can reconstruct both fire intensity and frequency, providing a clearer picture of how fire influenced pine forests. Certain species in the temperate forests are well adapted to survive and even thrive after fire. Longleaf pine (*Pinus palustris*), bur oak (*Quercus macropcarpa*), jack pine (*Pinus banksiana*), and ponderosa pine (*Pinus ponderosa*) are all well known for their resilience or dependence on fire systems to colonize new sites. Recent restoration efforts in the Southeast have concentrated on reintroducing surface fire to regenerate longleaf pine. The large swaths of jack pine "barrens" that exist today in upper Wisconsin and parts of northern Minnesota can be directly traced to the effects of fire on the landscape. Bur oak's bark is elegantly designed to withstand prairie fires while its seeds germinate in the burned grasslands. These species adapted to disturbance and thrived. As a result, ironically, the fire disturbance became their benefactor and not their enemy. The degradation was short term.

Wind, drought, and insects are also common drivers of ecological degradation, but they are more difficult to document. Consequently, researchers have less long-term evidence and understanding of how these disruptions have affected forested ecosystems. Recently, the immediate effects from large windstorms, including hurricanes, have offered a unique glimpse of the power of wind to reshape the landscape. The Harvard Forest in Massachusetts was the prime target of a hurricane in 1936. That hurricane changed the entire focus of research efforts there. More recent examples of extreme weather include hurricanes Katrina, Hugo, and Harvey, and the Boundary Waters windstorm of 1999 along the Canadian border. There, hundreds of thousands of acres of wilderness forest on the Canadian Shield were toppled by a storm and set off a chain reaction of further degradation. Eight years later, a 55,000-acre wildfire swept through the area, fueled by the windfall from that storm. This was followed by a simple lightning fire in a bog, accompanied by a drought, which blew up and burned through more than 100,000 acres of the same landscape. The smoke reached major metropolitan Chicago, 700 miles away.

These fires and the more recent conflagrations in the western mountains offer clues as to the domino effects of ecological disturbances.

One event often provides the trigger for another. Drought, wind, and fire often work together to change the course of forested landscapes and when human actions are factored in, the puzzle becomes more complex and often more destructive of both human property and forested landscapes alike.

For example, in the late nineteenth century, intensive logging in Michigan, Wisconsin, and Minnesota left hundreds of square miles of pine slash in the woods. It was unmarketable but volatile debris, and when the drought years of the early twentieth century hit, these cutover forests provided abundant fuel for lightning fires to flourish. The firestorms of Peshtigo, Wisconsin and Hinckley, Minnesota, which consumed thousands of acres, many small towns, and hundreds of human lives, are now etched into the history of the upper Midwest. The role of humans in driving these disturbances was painfully proven. From these devastating events came the birth of Smokey the Bear.

When we recognize that the forests are dynamic and not static environments, we place ourselves in a position to understand the role of disturbance and degradation in a focused light. Take drought, for example. As climatic change accelerates, drought has become a more important driver of degradation in many forested ecosystems, particularly the American Southwest. Moderate droughts affect seedling regeneration, insect populations, water levels, and stress hormones in trees. These variables are connected and codependent.

Because small seedlings require surface moisture for survival, an untimely drought may cut this process short and bring widespread mortality to both planted and naturally seeded sites. One forest manager learned this through experience on his first planting project 35 years ago. Of the 1,500 pine and spruce seedlings he and his crew planted, only 20% survived the first five years. After 10 years, 5% of the pine seedlings were alive. Spring drought was the principal culprit.

* * *

Drought also has a significant impact on insect populations. Drought, with its increased sunshine and higher daytime temperatures, gives bugs an expanded window of time in which to breed and multiply. Thus, many insect species produce additional sets of offspring (cohorts) during dry, hot summers. It is not hard to imagine the benefits to their population dynamics when beetles, sawflies, and moths breed more frequently. Their populations increase exponentially.

Because most insects are host-specific and are engineered to feed on one or two particular species of plants, certain plants suffer more than others. Recently, an outbreak of pine beetles (*Dendroctonus* spp.) has exploded in Ponderosa pine (*Pinus ponderosa*) and other species of the intermountain Rockies and turned alpine slopes into rust-colored graveyards. Further east, spruce budworms, whose diet is not limited solely to the spruce as is suggested by its name, feed on thousands of acres of balsam fir (*Abies balsamea*) in the boreal forest of the Lake States.

When climate conditions change rapidly, the established ecological system falls out of balance, and many insects are the immediate beneficiaries. During these events, it is not unusual for resource managers to get anxious calls from homeowners who have awakened to discover caterpillars seemingly falling out of the sky or woodpeckers chipping away at the siding of their houses. Suddenly there are so many more insects that the songbirds can't keep up with the abundant harvest.

Of course, not all disturbances in the landscape result in long-term degradation. We have already cited the benefits of surface fires for some species like bur oak and jack pine. In addition, the effects of moderate drought are usually short lived. Rain arrives eventually, and when it does, the local bird population expands. Insects are eaten or die from disease. Mother Nature restores the balance.

But severe disturbance has a different trajectory. Severe drought persisting for 10 years or longer permanently changes the forested landscape, sometimes replacing it with grassland. In the Black Hills of South Dakota there was a cycle of severe drought in the 1990s. Many of the pine forests burned. Where humans replanted, new forests resulted. On federal lands, however, the policy was to "let nature take its course," and it did. Today, thousands of acres of former forestland have succeeded to grassland. Today, national forestland in the Black Hills contains hundreds of square miles of recently established grasslands, easily visible from 30,000 feet in the air. Perhaps it is time to rename these lands as national grasslands in honor of disturbance ecology in action.

* * *

Unlike the 2017–18 wildfires in the west, most ecological disturbances are not front-page news. Often small disruptions go unnoticed. A mature white pine in the canopy is hit by lightning. Slowly its needles turn yellow and the woodpeckers move in. An old red oak falls on the forest floor, the victim of heart-rot and wind. Squirrels and salamanders set up a new home.

When these disruptions strike closer to human habitation, we take note. A beaver girdles favored willows along a driveway. A fire ignites from a careless cigarette, destroying a planted windbreak. A windstorm levels 30 acres of mature aspen and oak, leaving a chaotic jungle of dangling limbs. These moderate-scale disturbance events affect both the forest and the humans who reside there.

* * *

Twenty-five years ago this writer awoke one morning to find most of his mature hardwood forest flattened. Overnight an intense windstorm had churned it into a hundred acres of twisted limbs, upended roots, and scattered debris. This was a rude awakening to the angry power of natural disturbance. It was not an abstract philosophical state created by an idealistic notion of nature. It was a mile of splintered trees across a driveway, hundreds of uprooted oaks dying in the daylight, years of planting and pruning obliterated by three minutes of furious wind.

This dramatic event was also the start of a new trajectory, both for the forest and the resource manager. Here was an opportunity to digest disturbance ecology in motion. Why did it occur? What were the short-term ecological effects? How would the forest recover? What could I do to understand and aid the healing process?

What Restoration Is and Is Not

"Walk the walk, don't talk the talk."
—Anonymous

The process of forest restoration begins with an understanding of what it is *not*. This eliminates unnecessary confusion, illusions, and false starts. A path is cleared for better understanding and clarity.

For more than 25 years I have walked in northern forests with private landowners, making site inspections, taking forest measurements, and engaging in a Socratic dialogue. This dialogue begins with a host of questions, one of the first of which goes something like this: *"What are your long-term priorities and goals for the land?"*

The landowner's response informs my thoughts and influences my recommendations for management. It is not uncommon for the answer to the aforementioned question to include a profession to "do right by the land" or that the long-term goal is "sustainability" with a promise that the land not be abused. Most landowners want to believe that they are good stewards.

But when I hear these promises, I pause and listen more carefully, as we know that promises are easy to make but sometimes hard to keep. In one particular instance, a landowner should have held his tongue more closely. He called me to ask for an assessment and appraisal of his woodlands. When I arrived and we walked, he spoke repeatedly about wishing to treat his property as "sacred." I was intrigued by his use of such a strong word as "sacred." It soon became clear that the woodlot had not been considered sacred in the past. Past degradation was on full display. It was clear to me that it had been abused more than once in the past.

Among the lot's many abuses, there had been heavy logging conducted without the thought of forest health or long-term goals. Some of the cutting was old, but some of it was recent. This should have been a warning sign to me. There was also an area of fine young hardwoods, with a pole timber stand of well-formed northern red oak. As we walked, the landowner repeated that he wished to treat the land well. I couldn't help but think that perhaps he said it one too many times.

After a couple of hours on the site, I suggested that his woodlot was still young and recovering from past disturbances. At his request, I wrote him a follow–up letter and outlined that heavy harvesting of the good hardwoods should await further maturation, and that a light "improvement thinning" would be appropriate at the present time. While he seemed disappointed by my recommendations, he would not say more about his plans other than to reassure me of his good intentions. He never mentioned his need for income or his retirement plan.

Time passed, and I filed away his small invoice as "paid." One afternoon two years later, the phone rang. There was an attorney on the other end of the line. Usually, when an attorney calls it is not with good news, so I braced myself. He had a client who was unhappy with a recent timber harvest and was looking for help. This is not an uncommon request in my line of work. Unfortunately, many landowners have unpleasant harvesting experiences, and resource managers hear about most of them after the fact. By then, high-priced attorneys have usually been retained to mitigate the losses, if not in the woods, at least in the pocketbook.

When I requested more details about this attorney's client I was surprised to hear the name of the landowner whom I had visited two years earlier. Against my verbal and written advice, this former client had contracted his young woodlot to be cut, and it had been cut hard. Now he was unhappy with the result and was suing both the logger and another forester who had managed the harvest. As I listened to the attorney, my heart fell. Here was yet another nasty wound to a hardwood forest, and one that should not have occurred. After all, that land was "sacred."

After explaining this background, the attorney got to the point of his call. He was looking for an expert witness to back up the landowner's testimony of abuse. I paused and politely declined his offer, thinking to myself, "Why should I help a landowner who has already disregarded my advice?" In addition, I did not particularly enjoy the confrontation of courtroom disputes.

Surprisingly, the attorney then turned testy. He indicated that perhaps he would subpoena my testimony. This is a legal strong-arm tactic often used by aggressive attorneys. His purpose in mentioning the possibility of a subpoena was to force my hand, but in this case it was not going to work. I replied that I had advised my former client against a heavy harvest and indicated that I had put my counsel in writing. I asked if he would like to see a copy of the letter. Perhaps, I suggested, my testimony in the courtroom would not portray his client in the best possible light.

At these suggestions, the lawyer quickly dropped his aggressive tone and said he would get back to me. Not surprisingly, I have heard nothing from him since.

* * *

What causes landowners to ignore professional advice and to lose faith in long-term goals? Is it impatience? Is it greed? Or is it just that many landowners wish to control their own little piece of the planet by "doing their own thing" and then, when things go awry, try to justify their mistakes? Clearly, rhetoric and good intentions don't count. What counts is having patience, perseverance, and a clear plan.

The fact is that there are many forested landscapes that will not make the transition to something better. They may become a development, an agricultural field, or just another beat-up woodlot. These setbacks emphasize some of our challenges in the journey toward restoration. Renewing the health of forestland takes time and commitment, and requires that verbal promises be kept. For every model that works, there will be counterexamples of failure in which actions fall short of words. As we like to say in the woods, *"Walk the walk, don't talk the talk."*

* * *

Restoration cannot rely on empty promises to succeed. It is judged by results. With this in mind, it helps to start with a dictionary definition:
Restoration: ecology *the process of returning* **ecosystems** *or* **habitats** *to their original structure and* **species** *composition.*
(SAF Dictionary of Forestry, 2019)

Clearly, restoration is the result of an act or action. It is not passive, hands-off management. This is important, as it counters the preservationist theory that no management is the best management. Now that wildfires have swept through thousands of acres of public western forestlands, many of which were influenced by a hands-off approach, we need to understand the importance of positive human action. And we can learn from the long history of architectural restoration. Leaving the Pyramids or the Parthenon to decay would have been a form of gross negligence. Who would have benefited from this?

In this book I will argue extensively that leaving our forests without active management is a form of neglect. I call it *benign neglect*. What is needed is not less management, but a more thoughtful approach to what we do on the land. This includes lending an active hand as agents in the restoration process.

* * *

The primary prerequisite of restoration, whether of architecture or woodlands, is that degradation has occurred and health has been jeopardized (Gobster and Hull 2000). This degradation is measureable. A degraded forest is one whose value and values have diminished. The degradation may have occurred from improper harvesting, from an insect outbreak, or from a fire. In fact, the list of degradation agents is long and often includes multiple factors. Whatever the cause, the devaluation is measurable, whether in dollars, in water quality, or in loss of wildlife habitat.

Another prerequisite of restoration is that the degradation was caused by a disturbance. Disturbances are the catalysts for devaluation. There are a myriad of disturbances that can degrade a forest. Most of them are obvious. A few are not. For example, an invasive plant like buckthorn or kudzu can slowly spread over a forested landscape. The degradation may takes years to be noticed and the disturbance may be hard to discern. But both are present and the balance and health of the system is lost.

It should also be noted that not all disturbances degrade forests. Some actually improve them, and may be part of the restoration process. Take wildfire as an example. Most wildfires diminish value and health, but a prescribed burn in ponderosa pine or on an oak savanna, while clearly a disturbance, may be designed to benefit the long-term health of the system.

* * *

Finally, we need to recognize that forest restoration is both an art and a science. This synthesis is often overlooked by academics and supervisors. Creative approaches to returning value and health are as important as baseline data and accurate statistics. As resource managers and landowners, we will never have perfect information. This should not preclude us from trying new recipes once we have determined that serious degradation has occurred. And this is especially true in an era when climate change is adding unknown variables to future conditions.

Mother Nature is often our best ally in creative approaches to restoration. Why? Because she has been here before. She has restored forests on many old mining sites. She has restocked millions of acres of charred and cutover forests. She has turned sand and gravel into pine and palms. She knows soil conditions and competition, often far better than ecologists, hydrologists, and silviculturists. When we work with nature, when we employ the best science alongside creative observation, then we give our landscapes a better chance to heal. The pathway to successful restoration is cleared.

The Sky Has Not Fallen

"The sky is falling, the sky is falling."
—*From the fable of Chicken Little*

Often I am called to visit woodlands in counties near my home base in central Minnesota. When I depart from the office, the main highways are familiar and the road signs well known. Soon I turn onto less-traveled blacktop, county roads where the signposts are less familiar and the landmarks few and far between. Finally, I arrive at dusty township roads, where an unknown landscape greets me. It is usually not grand or imposing, but it is always refreshing and an adventure unfolds. On wheels and on foot, often with a landowner or forester at my side, we course the trails and the underbrush. Our footsteps compact soil that may never before have been touched by human heels.

Recently I was making such a trek with a couple of supervisors of township lands. At first we held to the trails as we discussed management history and options for the land. Then it was time to search out an old quarter corner survey marker to help determine property lines for the newly acquired parcel. I led the way off the trail and the going got rougher. The hazel brush thickened and slapped at our faces. We ducked under heavy spruce boughs and dogwood whips. My footsteps slowed and I could sense that one supervisor was unhappy with my bushwhacking course.

Finally, the brush lightened and we came out into an opening. As I paused to take a bearing, I heard the following words: *"That's the last time I follow a forester into the woods!"*

Fieldwork is a journey without the need for a hotel reservation, airfare, or knowledge of a foreign tongue. It makes the heart stronger and cleans the lungs. But it is not for everyone, even if you do find the quarter corner, as we did that day.

* * *

While working in the field during the past quarter-century I have witnessed many types of degradation on public and private lands. Early in my career, a visit to one of these locations would shock and sometimes depress me. A poor logging job is more than just an eyesore of bent and broken limbs and trunks. Regenerating species may be torn up. Half-rooted trees and stumps may litter the site. Ruts may fill the roads and logging trails. All of these are unnecessary burdens to the land and for the next generation (Fig. 1.2).

Figure 1.2: Recent logging site with residue.

Similarly, a woodland abused by heavy grazing is a testament to short-term thinking and usually economic depression. The remaining overstory trees have their roots exposed and often are only half alive. Invasive species like prickly ash and buckthorn are usually all that is left of an understory. And a blown-over forest is one that stuns the eyes with its random destruction, to say nothing of the financial loss it represents.

On my early inspections at these sites I focused on what had been lost. There were trampled hardwood saplings from a careless "skidder" operator. There were red oaks left on a pastured hillside dying from oak wilt. There was logging slash in the wetlands, or a burned-over pine plantation from a careless cigarette. Only the dead silhouettes remained for me to tally up the losses.

These graphic signs lingered in my brain long after I left the landing. A common pattern often marked the sequence of disruptive events: an absentee landowner, a goal of financial gain, an unmarked property line, a disreputable logger, a bad turn of the weather. The sequence read like a worn Hollywood script, particularly when it ended in litigation and I was called in to help judges and lawyers sort out losses. In many cases the damages could have been reduced or avoided with a few simple precautions: a paint line for the boundaries, a lock box for load tickets, a contract signed by all the parties to it, provisions for road repair, a professional manager overseeing the sale. But those necessities had been ignored and now people were angry. The lawyers were at the door.

As my whiskers have grayed, I have become more sanguine when called to a damaged site. Perhaps it is experience, or perhaps a tougher skin, but my eyes are no longer stunned. The damage is still there, sometimes in painful detail. But a different perspective rests on the horizon. Amidst the chaos and confusion, there is a prospect that the setback is temporary. What is the potential for recovery? Where is the silver lining? Therein lies my opportunity for restoration to begin.

* * *

I vividly recall the midnight winds on my land at Esden Lake. The calm black skies. The eerie silence. Then came the freight-train roar and the jarring sound of winds measuring more than 100 miles per hour. My cabin shook like a leaf on a tree, and a hard horizontal rain beat down. I was alone and the furies were obliterating the forest I loved.

That was 28 years ago. Could it have been that long ago, as I look out over a new forest that has risen from the debris? Today, the signs of rebirth have overwhelmed those of death and decay. Mother Nature has been steadily healing the wounds with the help of my human hands. Restoration has begun to take hold.

At first, on these blown-over woodlands, roadways were reopened and uprooted trees were limbed and piled. Sorted log piles grew on the landings with salvaged aspen, red oak, and red pine bolts. Semi trucks appeared with empty beds and disappeared with 10-cord loads destined for the nearest mills.

With the use of heavy equipment, debris was cleared or burned and the soils were plowed up and scarified. Brush piles burned brightly in the winter sky and the following spring a planting crew arrived with boxes marked "red pine 2-0" and "white spruce 2-2." Young bare-root and containerized seedlings were augured into the ground. More trails were cleared and seeded to clover.

Years passed as the pines outgrew the deer and the aspen sprouted for the sky. Brush saws returned, this time to thin overstocked young stands of pine, birch, and oak. White pine seedlings were bud capped or sprayed. And many cords of fuel wood have warmed my hearth on cool evenings. Today a young forest of paper birch (*Betula papyrifera*), red pine (*Pinus resinosa*), and quaking aspen (*Populus tremuloides*) greets my friends on the trails of Esden Lake. These saplings represent the next generation, and they will easily outlive my worn boots (Fig 1.3).

Figure 1.3: Esden Lake woods 22 years after windstorm.

This young forest did not spring forth overnight. There were times when its arrival seemed a distant dream: winters when porcupines got the best of some of the young pines; springs when the wind blew again and added more dead sentinels to the littered ground; summers when bark beetles seemed the principal beneficiaries of the rotting debris. But these events became part of the fabric in the recovery process, and offered me lessons in patience and adaptation. Today, I can even thank the winds at Esden Lake; they have taught me acceptance and the power of restoration in my own back yard.

* * *

Even with a helping hand, restoration takes time. Time for seedlings to establish their roots. Time before dead trunks decay into lignin or become a home for rodents and salamanders. Time for young saplings to find the light of the subcanopy. At Esden Lake these transformations have occurred in a quarter of a century. Today's visitors to the trails, including seasoned resource managers, remark on the beauty and health of the young forest. They do not see its traumatic past anymore. Instead, they see the future.

* * *

The healing power of nature is persistent. Beneath the surface, almost invisible to human eyes, plants recharge themselves. Small signs of movement appear. A thousand dark birch seeds rest on the snow. A single maple cotyledon sprouts from an old stump. Tamarack (*Larix laricina*) seedlings take hold in the grasses of an old log landing. And we are in a position to help them along.

Perhaps on your next visit to a familiar tract, try a visual experiment. Look for signs of a recent disturbance. The disruption may be only a broken limb or a rotting log. It may be a woodpecker cavity, an old stump, or a deer scrape. Or it may be more significant. A lightning struck pine silhouetting the sky. An elm skeleton with its bark stripped bare. A pocket of dead or dying red oaks. Whatever the agent, try to construct a story for the sequence of events. Is it a recent degradation or one that occurred many years in the past? Why did it happen? Were humans involved in any way? Are there ramifications for other plants nearby? Look closely for clues, and if you are with friends, engage them in your sleuthing.

Once you have created a background scenario for the event, look past the disturbance itself, beyond the decay and loss of life. What is changing on the site? Does more sunlight reach the forest floor? What plants are benefiting from these changes? Identify, if you can, the species of tree seedlings germinating in the duff. Attempt to reconstruct an ecological trajectory for the future of the site. If you are the landowner or land manager responsible for the woodland, consider methods to influence and expedite the recovery process.

Are more conifers a desirable option? If so, they will need help to overcome deer browse, drought, and shade. Are there acorns or oak seedlings amidst the litter? Perhaps white oak (*Quercus alba*) is a favored species and well adapted to the local native plant community. Does a new trail suggest itself now that the disturbance has passed? Where is the ribbon to mark it?

The opening chapter of restoration forestry begins with an attitude that favors change. The winds have passed. The fire and smoke have cleared. The chainsaws and wood processors have departed. A degraded forest is pregnant with potential. In the story, *Chicken Little*, our protagonist was mistaken when he dashed about warning his animal friends that the sky was falling down. In real life, the sky has not fallen. Opportunities lie ahead. The potential of restoration rests on the horizon.

CHAPTER II

A BACKGROUND TO RESTORATION FORESTRY

Picking Up the Pieces

Most adults have experienced times of severe disturbance in their lives. Perhaps it was the loss of a job or a divorce in the family. Perhaps a loved one was diagnosed with a terrible disease. Perhaps a fire or windstorm struck the neighborhood, causing loss of property and crushing dreams. These events have something in common: they are unexpected and alter the trajectory of life for individuals, families, and even entire communities. The victims of these events who are lucky have time to batten down the hatches or escape. But often we are caught off guard, shocked, and distressed when disaster strikes.

More than 10 years ago I was diagnosed with a congenital heart defect: a bicuspid aortic valve that resulted in an aneurism in my ascending aorta. Open-heart surgery was recommended as the first-line treatment. The realization that my life was on the line was not particularly reassuring. All of the best-laid plans for business and personal life were suddenly tossed into jeopardy. Home improvements were put on hold. My forestry business, 15 years in the making, went into suspension. A personal relationship fell by the wayside.

My first reaction to these events was shock and denial. It was time for a second opinion. Perhaps the operation was not necessary or could be postponed. But as time and extensive medical testing soon verified, this problem was "nontrivial" and my condition would not improve on its own. The aneurism was growing.

After months of hesitation, I finally took the plunge. The procedure was scheduled, as it turned out, none too soon. When I awoke from surgery 30 hours after the operation, my world was forever changed. For the next three months, I was lucky to hobble to the bathroom, brush my teeth, and return to bed, exhausted. My body seemed to resist all efforts to heal. I felt as if a freight train had paused on my chest. There was plenty of time to think about old age and mortality, and these dim specters loomed uncomfortably in the foreground.

* * *

When a forest burns, is cut, or blows down, a similar sequence of events takes place. There is an immediate shock to the system. Surgery is required. Firefighters, aerial tankers, wood processors, and salvage crews appear. Emergency teams are fully mobilized for the operating room. When the embers cool and the skies clear, the painful recovery process begins. Like open-heart surgery, the forest's healing is a long and slow process.

At Esden Lake, there were months when the blown-over woodlands resembled a war zone. For every fallen tree that was cleared, 10 more blocked the trails. For every acre of freshly scarified soil, five more lay in the rubble of roots, limbs, and brambles. The task of restoration appeared overwhelming and the results were slow to materialize. Why attempt this arduous reclamation effort? Wouldn't it be far simpler to pack up, leave the land and let nature take its course? And yet, I was a resource manager by profession. It was my responsibility to step up to the challenge of degradation in my own back yard (Fig. 2.1).

Figure 2.1: A blown-down forest.

When the shock began to wear off, there were moments of relief. Here and there, the immediate pain subsided and the sunlight reflected on the pale green aspen leaves. Then the first salvage crew arrived on site and the sound of chainsaws replaced the silence. They were a welcome sound.

When a forested landscape of any size has been degraded, the first step forward is to step back and make an assessment of the losses. This involves aerial mapping, site inspections, field measurements, and summarized reports. In many instances there are classifications of damage levels: light, moderate, or heavy. There are priority areas that require immediate attention: roads that need to be opened, structures that are damaged, colleagues to be consulted. In these early stages of recovery, patience is essential. Rome was not built in a day, as the saying goes; this same saying proves to be a handy reminder for those times when the recovery process has just begun.

There is an old adage in Hollywood about the five steps to success. The first step is PERSEVERANCE. Those who are patient and keep to the task at hand are rewarded. The second is your CONNECTIONS. In forestry, these are your colleagues, the salvage crew, the bulldozer operators. The third step is to be EASY TO WORK WITH. How hard do you push your crew? Can you smile on a bad day?

The fourth step to success in Hollywood surprises a lot of aspiring filmmakers, and it may surprise you too. It is TALENT or skill. You do not need talent at the top of the list for restoration or for filmmaking. You need perseverance, a good crew, and a positive attitude. Only then will your talents come into play. As old forest maps and models fade, be grateful for what is left behind and for the opportunity to begin again. And keep your eyes open for the fifth element of Hollywood success: TIMING or luck. It always helps to have luck on your side.

Ecological Roots

The roots of forest restoration begin with ecology. Why? Because ecologists have been studying restoration processes longer than other natural resource managers. This field began to build a record of restoration through the work of Aldo Leopold and others after World War II.

The term "ecological restoration" was coined by ecologist William Jordan. His early efforts at the University of Wisconsin arboretum led to his founding the first journal on ecological restoration in 1981. He focused on defining the "drivers" of degradation and their effects on a variety of species. Land conversion, exploitation, and increasing human populations were the leading drivers or causes. Secondary causes included invasive species, herbivores, and acid rain. Plants, invertebrates, vertebrates, hydrological functions, and landforms were all considered part of the complex of ecology and placed under the microscope. Ecological models were developed. The term "ecosystem" was employed and defined as a community of interacting organisms in the physical environment in which they live. Forests became a subset of this definition. "Habitat" was defined as the space, resources, and condition of species requirements to complete a life cycle.

From these early definitions and concepts the field of restoration ecology became established but, even in these early days there was not complete agreement of the goals or the role of humans on restoration. Some academics saw restoration as merely applied ecology, and hence derivative of basic, more important research. Others thought that restoration was misguided philosophically. Eric Katz, in the early 1990s, argued extensively that restoration was always false and created an "artificial nature." In his and other's view, nature was independent of human interaction and the best humans could do was to leave things wild. This philosophy helped strengthen the preservationist movement, the concept of wilderness areas and "leaving things alone."

But most ecologists embraced the role restoration could play the returning health to degraded ecosystems. From these roots, more focused efforts began to coalesce. Aldo Leopold motivated many to consider the role of wildlife restoration as a critical element of ecosystem health in his groundbreaking 1949 *Sand County Almanac*. The Audubon Society began with the snowy egret as it promoted returning keystone species from the brink of extinction. There was a new role for humans to play in the landscape and it was one of restoration. Whooping cranes, golden eagles, and spotted owls all became symbols of a lost landscape. Endangered species and habitat restoration became buzzwords of the nascent environmental movement. These efforts drove more interest in restoration techniques for forestland because many endangered species rely on large forested habitats for survival.

With the growth of restoration projects came more complexity. Why? Because natural systems have many more moving parts than masterpieces of art or architecture. For one thing, there is the new question of "original condition." While it is usually clear what the original condition of an old building or painting consisted of, the original condition of an ancient rainforest or an old oak savanna is unknown. Who makes the determination of what timeline is appropriate to consider? When all we have are estimates of pollen counts, carbon data, and a few other relics, can we ever know with precision what the original conditions of those systems were?

Other stumbling blocks also challenge some restoration efforts. The Chicago Restoration Controversy erupted in 1996 when local restorationists proposed cutting hundreds of acres of county forest preserves to reintroduce oak savannas and prairies to the Chicago metropolitan area. A political uproar continued for years, and finally resulted in a moratorium, largely because the public was not informed or involved in the early designs of the project. The idea of uprooting and destroying a half million trees in the Chicago metropolitan area struck a nerve and seemed counterintuitive to most of the population. Restoration took a black eye and a temporary back seat (Gobster 2000).

The Chicago Restoration Controversy served as a painful model and exposed many social issues of restoration that most scientists had ignored. Could restoration be used as an excuse to degrade a system? Was "meddling" with nature appropriate or was it domestication of something wild? In a democratic system, why was the affected population left out of the loop? The scientists were unprepared for these cultural questions with only their charts and data to guide them.

* * *

In any new field of inquiry or human endeavor there is never complete agreement on the best course of action. But disagreements may yield positive results in the long run. It is the Hegelian synthesis principle at work. In this book I make the assumption that forest restoration is an ecologically sound practice when approached with an understanding of its limitations. It is not a be-all or end-all solution. But I also disagree with those who argue that humans are not inherently part of the natural system and that there is an existential gap between artifacts of human origin and natural elements in the landscape. The sooner we understand and accept that we are part of the whole working system, and not some spiritually separate beings, the better the likelihood that humans will become constructive players in shaping a healthier, more resilient environment. Restoration is not a misguided attempt to dominate nature in a new way. It is instead a way to give back and rebalance our relationship with the world around us. It says to our critics that we are part of the solution, not merely the problem, and that our actions can make a difference both to the health of our future and to the health of the world we inhabit.

Reference Conditions for Restoration

"For now we see through a glass darkly; but then face to face."

I Corinthians 13:12

Restoration begins with a clear understanding of historic functions and conditions, whether in architecture or in forestry. This provides a baseline to prioritize goals and to set standards. Unfortunately, "original" or reference conditions in the forest are more difficult to determine than those in architecture or medicine.

For starters, nature plays an important role in the ecological restoration of forests. This does not occur in art, medicine, architecture, or other fields in which the baseline is a human construct. As a result, we are obliged to determine the timeline for the term "original." For example, there is a widespread notion that the original North American forests of pre-European settlement were untouched places, filled with ancient trees and endless biodiversity, a notion that these lands were virginal, peaceful, and serene, and did not feel the pulse of disturbance. But is this fact or fiction? Perhaps it is a romantic notion borrowed from nineteenth-century American literature.

What do we really know about presettlement forests? First of all, forest historians do not have a clear picture of temperate forests at any particular moment in prehistory. They have only rough sketches of what the landscape contained, based primarily on evidence from fire scars and fossilized lake pollen sediments. From these sketches, paleoecologists (those who study the ecology of the past) estimate that certain species, including Eastern white pine (*Pinus strobus*) and longleaf pine (*Pinus taeda*), were far more common in the past. Many pine species prefer xeric sites and are adapted to fire to help regenerate their seed. Hence, Eastern white pine tended to migrate from east to west and back again as the climate changed and moisture increased or decreased near the prairie border.

From early survey notes and lake sediment cores, it is also evident that aspen (*Populus* spp.) was a very common genus in the north and west and had been since the glaciers retreated into Canada. Because aspen is an early successional, or "pioneer" species, it requires full sunlight and mineral soil to regenerate. Those early conditions indicate that disturbance was frequent where aspen flourished. Fire, drought, and wind were common, more common than most early ecologists realized and reported.

What else have historians learned about presettlement American forests? They have verified that human influence was felt long before European explorers and trappers paddled the rivers of North America (MacCleery 2002). The journal entries of early surveyors and missionaries traveling the upper Mississippi Valley frequently record large fires and treeless hilltops in the unglaciated Driftless Area of Wisconsin, Iowa, and Minnesota. Local tribes used fire to drive game. These fires set back the forest and created large savannas of prairie grasses and bur oak knolls (*Quercus macrocarpa*). Where forests now thrive along the Mississippi River, open grasslands were common. With European settlement, the tribal fires ceased and oak savannas became rare. Currently the savannas in this region are among the most endangered ecosystems in the country and in need of restoration efforts.

In the few places where disturbance was rare, aspen was largely absent from the landscape. These protected areas, however, were the exception rather than the rule. In fact, most of the northern forests were filled with disturbance: fires set by lightning or tribes; windstorms, both large and small; droughts and climate change that moved the prairie boundary first west, then east, then west again.

Because fire records are easily verified by fire scars on old trees and charcoal sediments in the ground, we know more about the patterns and history of fire than the effects of wind, drought, and insects during the early postglacial times. However, recent large windstorms, including Hurricane Sandy along the East coast, Hurricane Katrina on the Gulf Coast, and the Canadian Boundary Waters windstorms, illustrate that severe wind is common and often precedes fire. Wind, in fact, is a critical influence on old forests and helps shape the patterns of early forest succession, just as it shapes forest preserves today.

In short, the landscape in presettlement times was far from peaceful. It was filled with disturbances that benefited some species and disfavored others. The pines were principal beneficiaries. The maples (*Acer* spp.) were not. Maples have thin bark and their cambiums are easily scorched by fire. Only where protected by rivers or lakes were maple forests dominant in times before European settlement in North America.

This picture of early American forests is vastly different from the view promulgated by a culture that prefers Thoreauvian myths of untouched nature to the reality of history. By deconstructing these myths and reconstructing history with more accurate information, land managers and policymakers can avoid misguided efforts to "turn back the clock" to conditions that may not have been common. But this is only a first step.

Reference conditions also assume that the climate and growing conditions in the future will mimic those of the past. With the increasing evidence of rapid climate changes occurring in temperate North American forests, this assumption may be invalid. Where does this leave current efforts for restoration?

Ecological restoration now recognizes that there was a range of natural variation in the presettlement forests (Cornett 2013). This emphasizes that no one particular baseline is appropriate for restoration efforts. Different managers and landowners have the opportunity to be flexible in how they prioritize their restoration efforts and which species they favor. By understanding the power of disturbance in the historical landscape, we prepare a more objective approach to reconstructing historical conditions.

"Sustainable" Confusion

Just as cultural myths influence modern attitudes toward forest management, language does as well, and words are usually not as simple as they sound. Let's take a closer look at a few of the terms that may muddy the waters.

One prime example of the difficulties of language emanates from the popular term: "sustainability." This is a term that reaches us from all directions: from the pages of journals, from media sound bites, from political analysts, and from advertisers. How does *sustainability* help us understand the role of healthy forests and move us in the direction of forest restoration?

Let's begin by looking at the word "sustain" itself, with its French root meaning "to hold up." The word also works to endure, support, prolong, and nourish.

A look through any dictionary shows that the word has many different and sometimes contradictory meanings. It offers to buoy a system as a life preserver or temporary life support. Is this how we wish to think of our efforts in the woods? Or perhaps the definition of this word is interested in merely prolonging the status quo? Is that what we want for our forest management activities? In the best of all its possible meanings, sustainability attempts to nourish us.

These are not idle differences, and they matter to forest practitioners. Even when sustainable forestry sounds like a worthy goal, we would be hard pressed to find widespread agreement regarding its fundamental purpose and meaning. Confusion and misunderstanding abound. I have seen the word used by economists and ecologists to justify opposite solutions to the same resource problem. I have seen politicians employ the word to describe short-term solutions to long-term challenges. What is our common goal? To buoy up? To nourish? To prolong? To legalize? For whom, and for how long? Isn't it time to find more precise and less ambiguous goals for efforts to improve the health of our forestlands?

* * *

In his groundbreaking book, *Basin and Range*, John McPhee explores the world of geology to his readers. He does this by telling engaging stories about geologic time. Geology is a sobering science in its perspective of time. Its clock is not measured in hundreds or even thousands of years. From the geologist's point of view, most forestry practices will never be sustainable. With plate tectonics and global weather change, many forests may be standing under water or buried in desert sands in the not too distant future. Only 12,000 years ago my woodlands in Minnesota and Wisconsin were nonexistent; they were part of a landscape covered in a thousand feet of ice moving down from the north.

Perhaps to give *sustainable* a chance, we should assume a time frame considerably briefer than that of what the geologist is concerned with. Still there are difficulties. Some resource managers believe that intensive plantation forestry is the most productive type of sustainable forestry for society. It is based upon a successful agricultural model adapted to the woods. Short-rotation loblolly pine (*Pinus taeda*), balsam fir (*Picea balsamea*), and red pine (*Pinus resinosa*) provide maximum fiber on a minimum of acres and exemplify how modern science increases productivity on a diminished land base. Successive generations of intensively grown, genetically improved growing stock sustain an industry that supplies much of the world's paper, cellulose, and plywood products. This approach is considered sustainable when the land is reforested or replanted after one growing cycle. But we don't yet know if the soils can support such an approach over the long run. Some ecologists and resource professionals remain skeptical of short-rotation pine, Douglas fir, and aspen. They note that research offers data from only one or two "rotations" (life cycles). They wince at the loss of native plant species in the understory. They question whether the soil can maintain its productivity for more than two or three generations, and cite European experience with coppicing to validate their concerns. Is this system really sustainable, or is it merely buoyed temporarily to support our consumptive needs?

What is the goal: To buoy or to endure?

Some forest managers are uncomfortable with short-rotation forestry, and advocate a different model of sustainability. They prefer to grow long-lived species of mixed-age classes with more species diversity. They place more effort on mimicking Mother Nature's hand, and use fire, natural regeneration, and uneven-aged management as tactical tools.

They do not prefer to plant or encourage the fastest-growing species to dominate. For these foresters, ecological factors as well as economic ones are the primary measures of sustainability (Dobbs and Ober 1995).

Perhaps both of these approaches are sustainable? (See Seymour RS and Hunter ML, New Forestry in Eastern Spruce, 1992.) But who makes the call about what is sustainable and what is not? I have yet to meet anyone in natural resource management who is opposed to sustainable forestry. It's the Midwestern expression of "motherhood and apple pie" all rolled into one. Loggers want to sustain their future income. Paper companies want to sustain a reliable source of low-cost fiber. Landowners want to sustain the productivity of their woodlands. Preservationists want to sustain the beauty of the land and their access to recreational enjoyment. Public agencies want to sustain a clean water supply. Each group has their set of standards by which to measure success. Rather than clarifying goals, sustainability takes on a veiled face. It is fluid and changes its demeanor based on whomever has the floor. Perhaps that is fair, and gives room for discussion and flexibility in forest management. But it also lacks clarity and may easily be misinterpreted.

* * *

Not long ago a group of landowners got together in the Midwest to form a forestry cooperative. They had long-term goals and named their organization "The Sustainable Woods Cooperative." The only trouble was that this cooperative struggled to make ends meet and survived for fewer than 10 years before it folded. Its name did not help the organization to survive; in fact, despite the name, the group was not creating a *sustainable* model.

Given these challenges, it is wise to acknowledge that there are other terms besides *sustainable* that are offered by those interested in returning forests to health. Rehabilitation is one such term. But rehabilitation is generally reserved for human habitations. The root of the word is Latin; it means to *dwell in*.

"Ecological" forestry is a third term that is offered by Jerry Franklin and others to sidestep the shortcomings of the word *sustainability*. But in my view, although ecological forestry focuses on understanding biological processes and silvicultural options of healthy forests, it largely turns its back on the two other legs that are critical to success: economics and cultural considerations. Is standing on one ecological leg better than standing on one economic leg? Or does it merely shift the weight to the other foot?

* * *

Because of all of these challenges with the current terminology, *restoration* is a term that deserves our attention. It has proven itself in art, in architecture, and in medicine; all disciplines with an illustrious past. Restoration of old paintings, historic buildings, and degraded forestlands breathes new life into valued resources. Restoration is not hands-off. It assumes there has been degradation and that renewal requires an active hand.

Forest restoration does not require a return to absolute (or imagined) historical conditions, for we have noted that original conditions contain natural variability and the climate may have changed. Just as a restored seventeenth-century building may contain air conditioning and plumbing, a restored forest may contain trails or even species that have adapted to changing conditions.

It is also possible, though challenging, to restore on a larger scale, at the landscape level. Wilderness areas that have been degraded by fire or wind are candidates for restoration. So are old mining sites and abandoned pastureland. Restoration can also be proactive to prevent degradation that is on the horizon. This is frequently the case in wetlands near agricultural fields where the risks of wetland loss are high.

Restoration projects require flexibility, planning, and monitoring. The flexibility offers an adaptive approach to changing climatic, economic, and social conditions. The planning includes cost–benefit analysis for high-end projects. The monitoring assures that goals are being met.

North American temperate forests have proven resilient in the past. But they have also been abused, both by humans and by nature's ways. Restoration provides an opportunity for professional land managers, landowners, and policy makers to participate in the healing process. It is a creative response to disturbance and degradation, and a new tool for land management challenges.

The Landscape Perspective: Beginning at Home

"Yonder stands the South Dome; a most noble rock, it seems full of thought, clothed with living light, no sense of dead stone about it, all spiritualized, neither heavy looking nor light, steadfast in serene strength, like a god."

—John Muir

John Muir was a landscape planner long before the profession as we know it today was born. As a visionary, he saw the big picture beyond the details. He drew pencil sketches of broad western landscape valleys like Yosemite as well as more detailed botanical studies of plants and leaves. He kept a journal of his wanderings and used it as a springboard for a larger mission to recognize and protect wild places in the American west (see Fig. 2.2).

Landscape planning need not occur only on a grand scale like the national parks in the high Sierras. Just as often it may be intimate, as intimate as a well thought-out design in your own back 40. The idea behind landscape planning is to create a more integrated and attractive space and to add diversity to the site. Where this design differs is that it usually stops at private property boundaries, but that does not alter its restoration goals.

The sites landscape designers in ecological restoration strive to integrate on a larger scale can include a watershed, a region, or even a biome. Sometimes the area of interest is a *political* or *economic* one, such as a township or a park. Sometimes it is an ecological one: a subwatershed or mountain ridge. And sometimes it is a *cultural* one, such as a popular recreation area along a river or lakeshore.

Each landscape planning project involves unique features that set it apart from its neighbors. Often the projects include geologic or human history. The Lake States, the Mississippi Valley, and the Big Woods,

Figure 2.2: Sierra Nevada, "Muir's Lake, 1st Merced, Yosemite Valley" Washburn Lake; undated. John Muir Papers, Holt-Atherton Special Collections, University of the Pacific Library. © 1984 Muir-Hanna Trust.

come to mind. In Tennessee it might be the Rim Plateau. In Oregon the Coastal Range. These are all ecological biomes dominated by similar geology, topography, and climate. The Driftless Area of southwestern Wisconsin and northeastern Iowa has deep loess soils, a hilly topography, and a lack of glacial debris. The Pine Barrens of southern New Jersey contains outwash glacial soils and forested species adapted to frequent forest fires. The Big Woods of central Minnesota is surrounded by rivers and lakes, and was protected from historical forest fires.

The rise of ecology, which studies the relationships between the parts of the ecosystem, has brought about the recognition that these landscapes each contain unique habitats and restoration goals for certain subsets and collections of species.

There are certain species that suffer in "old growth" or ancient forests. Jack pine, Karner blue butterflies, and sharp-tailed grouse thrive on outwash sands, a landscape of open grasses and shrubs, reinitiated by periodic fires. These species need full sunlight for survival. Other species, like black cherry, scarlet tanagers, and ginseng (*Panax* spp.) prefer seclusion and the shade of rich, moist soils. They find homes in the less disturbed hardwoods of the Upper Mississippi Valley or mountainous central Pennsylvania.

Many species of plants and animals are specialists. Over millennia, they have adapted to certain requirements, and they fare poorly when those systems are disrupted or fragmented. As the human population in much the United States has exploded over the past 200 years, many of these ecological landscapes have splintered or almost disappeared (Dobbs and Ober 1995). Not surprisingly, habitat specialists like Cerulean Warblers and Spruce Grouse have suffered from this disruption. At the same time, habitat generalists, those species that can adapt to the changes humans have wrought, have thrived from human influence. This includes many seagulls, ravens, and certain sparrows, along with migrants from foreign shores, including buckthorn, garlic mustard, eucalyptus, and kudzu. These species, like humans, are the current winners in the battle of landscape design.

* * *

Like the plants around us, humans also tend to live in landscape communities. We may choose to dwell in rural areas, small towns, or large metropolitan cities. Members of our communities share schools, libraries, banks, and supermarkets. We gather for coffee, for worship, or for block parties. Even if one is reclusive by nature, individuals are recognized by neighbors and counted in the census. In this way we are part of the natural community, not, as some would argue, separated from nature. To separate humans from nature is to put up a roadblock to integration and restoration. When we acknowledge that *homo sapiens* are part of the natural world, we free ourselves to concentrate on the issues at hand: the restoration of degraded landscapes where we live, work, and play. And to succeed in these efforts, we need to understand our forests as part of the ecological mosaic. Forested properties share attributes with neighboring forests and landscapes. The animals in the woods do not stop at your fence line or at the blacktop highway. Neither do the trees, plants, waters, or soils. All of us inhabit a larger labyrinth and are affected by what occurs nearby.

As a member of a cultural community, I have the wherewithal to influence neighbors. If I know the ecological history of the neighborhood, or how the water table is affected by nearby wetlands, I may share this information with friends and neighbors. If I am a hunter, I may discuss food plots or mature oak habitat. Or I may tap maple trees in the spring and bottle maple syrup for family and friends in distant urban centers.

All of these simple actions have a ripple effect in the landscape. Neighbors may watch my hardwood thinning and notice when a loaded logging truck pulls out of the driveway. Friends may spot a neighbor's name on a carving at the local gift shop. Community members may gather and tell stories at the local bar or café. Public officials may take note of a small article in the local paper. The word *restoration* spreads.

Landscape planning begins on a personal level with a few small steps. From there, the ripples spread outward. In this quiet, unassuming way, the activities we undertake on small or large woodlands have a greater impact. An example is set. Information is exchanged. Restoration takes hold. In the following three chapters, we will explore how ecology, economics, and the respective culture each of play a critical role in moving this model forward.

CHAPTER III

THE ROLE OF ECOLOGICAL
UNDERSTANDING

The Equilateral Triangle of Restoration

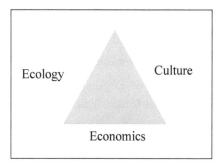

Figure 3.1: The equilateral triangle of restoration.

An equilateral triangle contains three sides of equal length that join to create a balanced geometric structure (Fig. 3.1). Similarly, a comprehensive approach to forest restoration is tripartite, incorporating elements of culture, ecology, and economics. It is a three-legged stool.

Unfortunately, the historic use of forestland by humans has been focused primarily upon only one of these legs: the economic one. The ecological and social components have been relegated to the back seat. Herein resides an ancient problem. Civilizations from Mesopotamia to Spain, from ancient China to early America, have relied upon forestlands as an economic resource. Without exception, these cultures have exploited woodlands for fuel, for cropland, and for wood products.

The long-term results have been reoccurring cycles of advance and decline, as John Perlin eloquently outlines in his book *A Forest Journey*. Perlin notes that as succeeding civilizations rise in power and influence, their woodlands are tapped and exploited. When the local resources are depleted or exhausted, the culture usually turns to distant shores for fuel and fiber. Ultimately, there are no longer adequate resources within reach of the growing demands. The civilization then struggles to maintain itself and subsequently spirals into decline. For those who study forest history, the correlation between the health of the forest and the health of the associated human population is striking.

Of course, there are myriad factors involved in the decline of Persia, Greece, Rome, Spain, and, more recently, the United Kingdom. But the fact remains that as these great empires grew powerful, they each exploited their mature natural resources and ultimately paid a steep price for the demise of them. With this somber history in mind, it is natural to wonder whether we in the United States will repeat the same deep-rooted mistakes and pay the price ourselves. In our current consumptive and debt-ridden ways, the United States continues to edge closer to a dangerous cliff. What's more, it is not merely our own natural resources that are at risk, because we mine the finite resources around the world to support our global economic engine.

The philosopher George Santayana reminds readers to take heed of the past with his famous dictum, "Those who do not remember the past are condemned to repeat it." How can resource managers and landowners respond in the years ahead to assure our forestland's health and vitality? Is there an opportunity to resist the temptation of short-term gains in favor of long-term goals?

Fortunately, a few positive signs have appeared. For starters, the health of the environment has been recognized as a critical element for healthy societies. The role that well managed and wild forestlands play in providing clean water and clean air is historically and scientifically undisputed. As a result, thousands of acres of forestlands in watersheds near many of our urban centers are protected for just this purpose. In addition, the attributes of mature forests have recently been studied for their contributions to genetic biodiversity and species complexity, to say nothing of their aesthetic and recreational benefits. In the research community, scientists today measure the interrelationships between the climatic functions and changes in flora and fauna. Biologists and pathologists search for the secrets of plant diseases, and hydrologists measure mechanisms for producing cleaner water from rivers and aquifers. All of these researchers conduct sampling in the forests alongside foresters in cruising vests. Clearly the role of ecological understanding of forest functions is on the rise.

* * *

The third side of the equilateral triangle is culture; all forests, even those in the taiga or jungle, are affected by human society. The influence of culture, however, is more difficult to measure than that of air or water. In addition, there has long existed cultural tension between the needs of rural communities and the health of the forests they depend on.

The woods have been a draw for people for centuries, offering respite in addition to resources. Thousands of years ago, members of early cultures journeyed to the woods to hunt, gather fuel wood, and perform ceremonies. They took spears, blowguns, and canoes. Today, we walk in forests that have been cut, burned, pastured, or blown-over, perhaps multiple times. And we bring modern amenities: propane stoves, GPS systems, and digital cameras along with our campers and tents. Despite these superficial differences, many of our goals are similar to those of our ancestors. We join friends and family for a fall hunt. We pause on a trail to enjoy magnificent autumnal colors. We ski the slopes of a favorite mountainside or canoe in backcountry waters. Perhaps a campfire warms us in the cool evening shadows.

In the United States today, towns have risen at the border of national parks and along many rivers where paper mills garner their power. Retirees have settled on mountainsides where logging trucks are common. Hikers still marvel at ancient cave art and wall etchings carved into the sides of cliffs. Our cultural history is born, reborn, and layered with that of the next generation, and placed alongside the histories we have inherited from the generations that came before.

The journey to restore America's forests requires all three sides of the equilateral triangle to collaborate in order to succeed. If one side is neglected or damaged, the whole system suffers. In the following chapters we will explore each of these individual legs of the triangle in more detail. What specific roles do *ecology*, *economics*, and *culture* play in establishing restoration priorities and avoiding pitfalls?

The Carbon Cycle in Temperate Forests

Forest restoration begins with a clear understanding of water and soil. These are the first building blocks of terrestrial growth. Add a moderate climate, adequate sunlight, and a seed source, and you have the foundation for a forest to find and build a home.

Trees are remarkable ecological factories. They process sunlight, water, carbon dioxide, calcium, and more than 20 additional biological building blocks, including phosphorus and potassium, through their systems. They run these raw materials through a complex production line, first by using sunlight to break down the carbon dioxide and water molecules. Finally, they recombine these components to produce a woody plant of cellulose and lignin that rises higher than most other plants on the planet. From start to finish, it's quite an operation. Furthering this remarkable process, they also produce much of the oxygen we mammals breathe.

Trees manufacture oxygen by converting plentiful carbon dioxide molecules into two separate elements: oxygen and carbon. The oxygen is a byproduct of this process and most of it is released into the atmosphere. This also makes trees a tremendous source of reserve carbon because the bulk of the resulting carbon is stored in their roots and trunks. The oxygen is released into the atmosphere, making human life and life for thousands of other species possible (Fig 3.2).

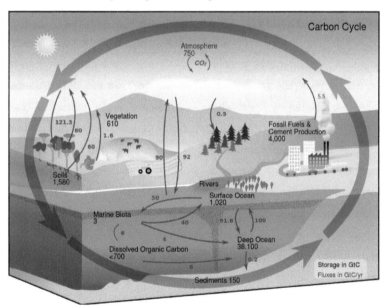

Figure 3.2: The carbon cycle

Not only do trees offer oxygen and sequester carbon, but further still, they produce another by-product we use for shelter, transportation, and heat: lignin fibers. Even in the darkness of night, the stomata on the needles of conifers perform this exchange of oxygen and atmospheric carbon dioxide. Every day, we breathe the results of the forest's labor. Common sense dictates that we should reward the trees and those who grow them because their ecological functions are critical to our survival. Unfortunately, however, rewarding the trees themselves is more easily said than done.

* * *

It is widely recognized in the scientific community that sequestration of carbon by living plants is an important factor in limiting the buildup of carbon dioxide in the atmosphere. Unfortunately, there is less agreement, either politically or practically, on how to manage increasing levels of carbon dioxide. As a result, there are conflicting international standards, protocols, and directives as countries, industries, and activists weigh in on the problem and jockey for the inside track. In the United States, there are conflicting approaches. One centers on the merits of "cap and trade," a system in which companies that emit significant carbon through the course of conducting business (namely, energy and utilities companies and manufacturers of goods) purchase carbon credits from forestland holders to support efforts related to storing carbon. In addition, "allowances" are sold through auctions and the revenues are used to reduce greenhouse gas emissions and improve public health (Bubser and Poillet, personal communication with the author, 2019). Although purchasing and trading carbon credits sounds straightforward, the challenges to this approach are both technical and political. Technically, it is difficult to measure all carbon inputs and outputs with accuracy. In addition, those on another side of the political debate hold that climate change is natural and that establishing a cap and trade system is complex, expensive, and unnecessary.

A third approach to the carbon conundrum focuses on a new "carbon tax" as a means to encourage companies to reduce their carbon emissions. This approach penalizes companies that emit a high level of carbon, but does not reward those who produce clean energy or establish emission reduction targets. The principal advantage of this strategy is that the government reaps new taxes without having to establish the bureaucracy necessary for a carbon credit program. Hence, it is a far simpler plan to implement and monitor.

Many managers and the overwhelming majority of scientists view increasing global temperatures and rising greenhouse gas levels as early warning signs that the atmosphere is changing at a rate that threatens human health. They urge the legislators and regulators to act promptly and mandate new mechanisms to slow carbon dioxide emissions. Only time will tell if they are heard or heeded.

Although all sides in the carbon discussion appear to recognize the high value of forests and trees, they differ in which values are most important and how active a role government should play in regulatory controls. If history can predict the outcome, it is likely that compromise will be reached. In the meantime, there is a current political deadlock and the clock is ticking. Carbon dioxide levels continue to increase in the atmosphere, the polar ice caps continue to shrink and melt, and ocean levels continue to rise.

* * *

As forest practitioners, we need to understand the carbon cycle and its potential to influence the atmosphere through various forest management activities. For example, it is easy to understand how carbon is stored in the products from harvested trees. The carbon is locked up in our houses, our furniture, and our books. On the other hand, when we burn wood or when a forest fire consumes a swath of forestland, tons of carbon is released into the atmosphere. Ecological understanding recognizes that forests are more than sources of timber and that trees reward us in multiple ways. The reduction of carbon dioxide and the production of oxygen are two important ecological benefits to all of society. There is a good reason why our lungs feel better on a spring walk in the woods: our oxygen level is higher as a result, and the effects are immediate and obvious.

Natural Regeneration and Soils

"Don't treat the soil like dirt."

—Anonymous

The United States put a man on the moon in 1969. That effort took significant technical expertise, and the result was impressive. But despite these impressive skills, it is time to acknowledge the limitations of our scientific knowledge. For example, we still do not understand many reproduction and regeneration aspects of most tree species in the temperate forests. We have been taking them for granted for a long time.

In the Lake States, for example, a high percentage of the forests were affected by harvesting white pine in the late nineteenth century. When industry moved west, some cutover lands were cleared and farmed. Many more acres were burned and then abandoned. This is an old pattern of human exploitation. New Englanders demonstrated the same behaviors in the 1700s with the pines and white oaks of their woodlands. The British exploited their forests in the 1500s to fuel charcoal and iron production and support their growing economy. The Spanish did the same in the 1300s in their quest for ships and riches (Perlin 1989).

In North America, we have been lucky thus far. Many forests have come back after earlier challenges of deforestation. Quietly, without government subsidies, nature has reforested large portions of the North American continent on her own. When American agriculture moved west to more fertile soils, the old rocky pastures and fields of New

England filled again with white pine and oak. The hillsides of northern Appalachia regenerated with hickory (*Carya* spp.), black cherry (*Prunus serotina*), and tulip trees (*Liriodendron tulipifera*). The tribal burned hillsides overlooking the Driftless area near the Mississippi River came back to support oak and other hardwoods. While they were never quite the same, this leads us to some important questions. How did these forests return? Why did certain species survive and others fade? What were the mechanisms of reproduction and survival?

Because restoration ecology is the art and science of understanding natural processes and interrelationships, these are questions that ecologists ask when studying regeneration patterns. What adaptive circumstances are effective to restore preferred species, both at the site and landscape scale?

In order to address the influences of habitat, moisture, and light, we need to start at the beginning, with the soil: the dirt and gravel at our feet. In northern forests, glaciers heavily influenced the soils and forest topography (Fig. 3.3). In some areas glacial till still abounds, formed as

Figure 3.3: A glacial till. Photo from USGS.

the colossal forces of the ice churned up the ground at its feet. Glacial till contains a tremendous variety of soil types and nutrients, all roughed up, turned over, and deposited. Not surprisingly, it is rich with nutrients and supports a wide variety of plant life. In other areas, the glaciers and their melting streams left the landscape with outwash sand and gravel. The glacial rivers washed away most of the soil's smaller particles (the silts and clays), leaving the larger particles in their wake. These *xeric sites*, the technical term for ecological dryness, were prone to drought. As a result, fire was a common occurrence. The pine barrens of Wisconsin and New Jersey are two present-day examples of xeric landscapes, as are the sands of Florida. Many species of pine are best adapted to these landscapes because they are more tolerant of drought and many require fire to release their seeds from their hard casings. Frequent wildfires also eliminate many thin-skinned cousins like the maples.

The northern landscape also contains terminal moraines, another glacial byproduct, scattered over the postglacial debris. These hilly pockets announce where a glacier paused and dropped some of its heavy load. The soils on these moraines tend to contain more fine-textured materials, loams, and clay. As a result, terminal moraines hold water longer than the outwash plains. In the language of hydrology, they have higher moisture-holding capacity and are better suited for hardwoods, which are demanding of nutrients. A terminal moraine produces high-quality oaks, sugar maple (*Acer saccharum*), and basswood (*Tilia Americana*). Unlike their more-frugal conifer cousins (the gymnosperms), temperate hardwoods lose their nutrient rich leaves every year. Those beautiful fall colors end up depositing another layer of calcium, potassium, and other minerals on the soil surface. As they slowly decay, the nutrients leach back into the soil and to neighboring root systems. It is an ecological loop.

Glacial history has shaped the form, texture, and structure of the subsoil on each of the topographic landscapes in the northern forests, and it is the quality of the subsoil that determines which elements are abundant for growing plants and trees. These soils also determine the amount of moisture that is held near the surface and the available organic material, both so critical for seedling germination. Variable soil conditions, including till, outwash, and moraine, remind me of a childhood story: *Goldilocks and the Three Bears*. Is the porridge too hot, too cold, or just right? Most plant species favor loam over sand or clay. They prefer their porridge just right.

* * *

Moisture, most commonly in the form of rain, is another key to ecological health and restoration opportunities in the forest. Approximately 25 inches of rain annually is the magic number for temperate forests. This measurement stands as the dividing point between forestland and grassland. Where rainfall drops to less than 25 inches per year, the eastern North American forests fade and the Great Plains begin. On the eastern side of this invisible boundary rise oak, pine, and poplar forests. To the west, the arboreal presence rapidly disappears, to survive only where moisture is abundant, such as along the sides of streams, gullies, and in higher elevations. The old prairie grasslands, today transformed into wheat, corn, and grazing land, stretch into the horizon. It is relevant to note that these invisible meteorological lines have migrated many hundreds of miles eastward and westward over the centuries, as the continental climate has shifted (Pielou 1988). Pollen counts in lake sediments verify this pattern of vegetative migration.

It is also not surprising that when moisture and temperatures increase, so does species diversity. The greatest species diversity in the forests of the United States occurs on the moisture-laden slopes of the southern Appalachian Mountains, where winter temperatures are moderate and more than 75 inches of rain falls in a single year. This combination breeds a plethora of floral life, including more than 200 species of trees and many more shrubs. By contrast, on the edge of the northern plains, where winters are cold and moisture less bountiful, fewer than 25 tree species are present today.

Just as humans build most cities where water is plentiful, trees and shrubs put down roots on sites where water is available near the surface. Because outwash sands do not hold water well, only a few species survive there. Blueberry (*Vaccinium* spp.), jack pine (*Pinus banksiana*), and pitch pine (*Pinus rigida*) are all well adapted to these xeric sites. On the other hand, a terminal moraine contains a thick understory and the competition for water and light is intense. Sugar maple, cherry (*Prunus* spp.), elm (*Ulmus* spp.), red oak, butternut, dogwood (*Cornus* spp.), leatherwood (*Dirca palustris*), and a host of forbs greet the eye. In a mountainous terrain, the mix of species may vary within merely a few paces. In the middle of a ridge, hardy pines or oaks may dominate. On the southern flanks, where the summer sun is strongest, drought-tolerant grasses and red cedar (*Juniperus virginiana*) may appear. And on northern slopes, where it is cool and moist, boreal species such as birch (*Betula* spp.), green ash, and quaking aspen find a home.

Most of the forests in the northern and central United States have regenerated following degradation without human help. Mother Nature has adapted to grow trees in unlikely places: on rocky ledges or old mining spoils. She produces trees in wetlands and on sand. She knows how to regenerate trees following windstorms, fires, and clear-cut logging. This is an impressive display of ecological know-how, worthy of study and emulation. Certainly we can do worse than observe Nature's habits and mimic her clever restoration strategies.

* * *

Every spring when I was a child, I enjoyed planting season in my grandmother's garden. As the days grew longer, it was satisfying to labor over a plot of dark earth, with its scent of humus and its mottled textures. With mud-caked hands, I anticipated the future of the small seeds dropped into narrow slits in the earth. It seemed a minor miracle when those first wisps of green pushed out of the soil and opened small leaflets toward the light. These memories are refreshed today when planting season arrives and I stand in a forest opening, expectant, like a child. After more than 60 years, I have learned that it is not how many trees I plant that matters. It is how many seedlings survive the first five years in the ground. There will be challenges from drought, from competition, and from deer. Many of these offspring will not make it to adulthood. But if I measure the soil quality before planting, I should be capable of matching the species to the site, as Mother Nature does. This will increase the odds of success, a first step in ecological restoration. For those that do survive, the new generation holds promise.

The Marriage of Water and Trees

"Drip drop drip drop drop drop drop / But there is no water"

—T.S. Eliot: The Wasteland

The words of T.S. Eliot's epic poem resonate in the kitchen. They remind me what I take for granted when turning on the faucet. What happens if the rain ceases and the aquifers run dry? What follows when freshwater disappears and surface streams shrivel? This is not a farfetched thought; it is a fearful prospect in the American Southwest and elsewhere.

Imagine a landscape that is barren and frigid, where shades of gray dominate the retina and the wind moves in silence over a treeless topography. Lifeless hummocks, hills, and valleys recede into the distance in vast stationary undulations. No water is visible or audible. Arid gulches extend in all directions, bleached white by the sunlight and salt. Rocks protrude in jagged shapes and a high plateau stretches to the horizon.

This is not my imagination. It is an ecosystem of the high desert in the Peruvian Andes, one of the driest and least habitable landscapes on earth. Less than five inches of rain falls annually on this plateau, and yet it adjoins the high Andes to the country's east. There the snowmelt from the eastern slopes rushes in torrents toward a vast rainforest, down the Urubamba and Apurimac rivers toward the Amazon basin.

The treeless western Andean high plateau is a barren reminder of the power of water. On one side of the 20,000-foot peaks rests a deathbed of powder and rock. On the other stretches the most prolific, complex rainforest in the world: the Amazon Rainforest, home to more than 14,000 tree and shrub species. The essential difference is not latitude, geography, or cultural history; it resides in a four-letter word: rain.

Water and forests belong together. Theirs is a long-term relationship, inextricably entwined and codependent. Their voices coalesce where moisture condenses in the clouds at high altitudes. There, the headwaters of great river systems are born: the Amazon in the Andes, and the Ganges in the Himalayas. In North America, both the Missouri and the Columbia river systems originate in the northern Rockies. The combination of moisture and mountains breeds many of our most spectacular forests. Where moisture is lacking, the forests are as absent as rain clouds. Closer to my home, the Great Lakes of North America harbor 84% of the surface fresh water of the continent. That's a lot of water, enough to spread it out a foot and a half deep over the entire continental United States. As other regions of the world suffer under wilting sun and receding fresh water tables, the Great Lakes enjoy an unprecedented wealth of water, and it is stored not only in the lakes of the region, but also in millions of acres of wetlands, in thousands of streams, and in deep underground aquifers. It is this water that also replenishes the headwaters of the most famous river system in North America, the Mississippi River.

The Mississippi River begins its journey in northern Minnesota, practically my back yard. It starts out as a narrow creek, trickling from a series of small lakes. It flows north, then east, and finally turns south toward the heartland and ultimately the Gulf Coast. As it meanders from its headwaters, the Mississippi is fed by neighboring wetlands, streams, and larger lakes. For the first two hundred miles, these headwaters intersect with a northern forest, a forest that survives on snow and rain. Some of the moisture that the forest absorbs vaporizes into the atmosphere and coalesces as clouds. It is later recycled as rain. Some of the moisture seeps into the ground, saturating the soil and seeking openings to nearby wetlands or lakes. It is this cycle that provides the breeding ground for the forest, a breeding ground of glacial till, a moderating climate, and water. Because it is my home, I have come to watch its cycles carefully. Many of the personal stories that follow occurred in this landscape of green and form the roots of my active hand in restoration.

* * *

Conifer needles come into focus in the foreground. A gentle wind awakens the eardrums as raindrops coalesce. At first they appear as a soft mist, a few scattered grains of moisture on chapped forearms. Then the sky darkens and a chill reaches the chest.

I pause before seeking shelter, first replacing my tatum in its vest pocket. The drops on my forehead are my friends, the lifeblood of the forest that surrounds me. This forest would not be here if not for the moisture on my cap. I recall the cold stone mountains of the western Andes and the dry high plateau of the American Southwest, landscapes I have seen with my own eyes. Not long ago, the forest I walk today was only a dream, a concept buried beneath a thousand feet of glacial ice. The ice had buried the landscape, but the dream of a future forest lay dormant beneath the surface.

The Birds and Bees

What do I know about birds? To be honest, not very much. After all, I am a forester, not an ornithologist. Ask me about lignin or *anthracnose* or Ips beetles, and I will respond with ease. But birds? Except for the occasional phoebe or familiar "hoot," I know little more of birds than the joy of their great cacophony that welcomes spring.

Fortunately, some of my clients are more knowledgeable about birds than is their forester. My first experience listening more closely occurred more than 25 years ago near my land at Esden Lake, while conducting a field inspection for a neighbor. His woods had been hit hard with the same windstorm that woke me from a deep sleep. As we discussed the storm and salvage options, suddenly he stopped talking about the storm and pointed skyward.

"That's the male. He knows we're here."

I looked up. A single bald eagle, its white head unmistakable against the blue sky, circled overhead. The tables had turned; I was no longer the observer, but the observed. Quietly, my neighbor pointed to a small clump of mature white pine about 200 yards to the north. There, in the canopy, a dark silhouette outlined the sky. Eagles and white pine go together in the north woods, and that eagle had an active nest on the property. It was no wonder the male was watching us closely.

We paused on our walk and my neighbor's voice turned childlike and excited. The eagles had returned, he informed me, and the female was sitting on eggs. The tone of his voice offered a key to his enthusiasm and concern. I took a mental note. Whenever landowners get a glint in their eye or speak in reverent tones, they offer a portal to their priorities. My job is made easier. It becomes one of translating. The recommendations I make merely provide a mechanism to enact goals that stem from the landowner's passion for their land. The forest and the caretaker both benefit. For this landowner, the restoration of nesting habitat for the eagles was clearly a top priority (Fig. 3.4).

Figure 3.4: A Bald Eagle.

Once my neighbor had spoken about the eagles, the remainder of that morning was spent less focused on trees and more on birds. While I still took down data and notes about his forest cover types that had been damaged in the storm path, we focused on the nesting eagles. He knew that they were a threatened species, and had called the Department of Natural Resources (DNR) to report them. They sent a non-game expert to visit the site and record it on a GPS database. She gave the landowner some information on how to identify signs of a successful hatch. They discussed how to protect the area from human disturbance, including any salvage harvesting that he might choose to conduct. As a result, the landowner decided there would not be any operations in the spring or early summer.

Over the years, I have kept in touch with my neighbor. Every year, in late winter, he searches the sky for signs of the eagles' return. One year they were absent, and he thought the nest was abandoned. The following year, two birds returned. Were they the same birds or the offspring? He could not answer, but he did know more about eagles and their nesting habits, and his forester was on a learning curve as well. The protection of the nesting site became a top priority for him and his family.

* * *

Since that day I have experienced a number of remarkable bird encounters in the woods. One landowner and his wife took me for a walk and began to identify all the nests along the path by the species who built and had occupied them. They regaled me with their knowledge: "That one's a red-eye vireo's nest," and a few steps later, "A yellowthroat lives there."

Many of the nests look similar to me. But they were not the same, and it took an experienced eye to note the differences. The nests were at a variety of heights from the ground and were constructed of different materials. These birders know the details and enjoyed sharing their knowledge with the uninitiated. Once again, the tables were turned on the professional resource manager at their side. I was the student, not the teacher, and sometimes that can open your eyes to new management options.

On another occasion a landowner at my side suddenly began to mimic the lone cry of a hawk as we hiked in his woods. I stopped short in my tracks when a similar call was returned from high above. A conversation between the two followed. I listened in disbelief as the two males marked their territory. Only one of them was a red-shouldered hawk with a nest nearby.

Sometimes my colleagues impress me with their knowledge of birds. On a field tour with fellow professionals, an urban resource policy analyst, who spent most of his time in the halls of the state capital, calmly identified five different warblers from the songs emanating in the thick hazel. To my ear the songs all seemed to blend and all I saw was a wall of green brush and an occasional flash of banded wings. He, however, could distinguish the warbling variations that poured forth in cascading rhythms from the green jungle surrounding us. It was mating season and the warblers were putting on a show for their mates and anyone else who happened to be nearby.

These birders proved that like most skills, knowledge of avians is an acquired one. Because I am trained to focus on trees, I see selectively. Only certain components of the forest and its structure stand out to my eyes. Most of the forest aviators are absent on my radar. And they have been living in the woods for a long time.

* * *

Lately, I have begun to pay more attention to bird habitats, particularly tree cavities. Many birds are cavity-habitat specialists, with strict requirements for where they feed and build their homes (MacCleery 2002). Pileated woodpeckers, with their long wingspans, prefer open, older forests or those disrupted by storms and aging aspen is one of their favorite species. There, amidst the conks and Hypoxylon cankers, they search for clustered populations of insect pupae. When you spot a large cavity in an aspen, it was most likely created by a pileated woodpecker. This provides a clue of the local forest structure and history (Fig. 3.5).

In an urban setting, one of the telltale signs of the invasive emerald ash borer (EAB) beetle is an increase in woodpecker feeding in the high crotches of ash trees. It is there that the beetles first migrate. Because woodpeckers depend upon the beetles for sustenance, they often find the insects long before arborists or entomologists do.

Some bird species dislike disturbance and prefer seclusion. Each has a habitat niche where it breeds and feeds. Cerulean warblers nest on undisturbed hillsides of hardwood forests near streambeds. Red-shouldered hawks require large acres of mature oak forestland. Ruffed

Figure 3.5: Woodpecker cavities.

grouse migrate to thick young aspen stands. Redheaded woodpeckers inhabit open-grown, savannah-like, woodlands.

Other birds prefer disturbance. Pileated woodpeckers are one example. Their population increased dramatically following the windstorm on my own land at Esden Lake. So did that of the flickers. And not all warblers like mature habitats. Golden-winged warblers prefer immature aspen and open-grown brush.

Like humans, some birds like it hot and some like it cold. Great gray owls swoop into my neighborhood during severe snowy winters when the populations of voles and mice are depleted on the Canadian Shield. Red tanagers appear in the northern forest only during the warmest months of summer to breed when the heavy foliage obscures their brilliant plumage. They, like hummingbirds, spend most of their lives in tropical forests, where insects and nectar are plentiful year round.

Some birds exhibit a plethora of habitat requirements, and many suffer when the human population increases and the birds' habitat is divided into smaller woodland parcels. Wood thrushes, redheaded woodpeckers, and cerulean warblers have all lost portions of their habitats in North America, and their populations have diminished recently as a result. But some of their distant relatives, particularly the scavengers, thrive when the human population increases. Blacktop highways are sprinkled with crows mopping up road kill. Landfills are thick with herring gulls pecking out leftover meals. Farmers' barns are filled with sparrows and phoebes. These species benefit from human ways and partake of the bounty of human debris.

* * *

Whether it is for the birds, wolves, or butterflies, habitat protection is a critical arm of forest restoration, and knowledge of the ecological parts helps us to make better-informed decisions. Just as a canary gives warning to unhealthy working conditions in mines, avian populations and behaviors can provide an early warning system for forest health. Rachael Carson demonstrated that eloquently in her now-classic *Silent Spring*.

Birds are also connected to a deeper, prehistoric past. The robins of spring, the bats of summer, the grouse of autumn: each have evolved with the forest and hold a place in the ecological web. Restoration recognizes and strengthens this web. It is an opportunity that beacons many rewards.

Forbs as Indicators and Allies

Foresters are trained to pay more attention to the trees than to birds or small plants. For example, how do we determine when a woodland is healthy? What are the signposts of vitality? Where do we look for clues? Some of the answers are obvious and the forester's first impulse is to study the overstory. How long are the new leaders on the pines in late spring? How much Hypoxylon canker blackens the trunk of a trembling aspen? Are there signs of oak wilt or anthracnose on the red oak leaves? How rapidly is the bole of birch expanding? These simple clues are observed daily in fieldwork and recorded in tally notes. To the careful resource manager they are smoke signals, harbingers of what may be around the corner, and early signs of disease, drought, or stress.

As a result of my focus on the overstory, looking up comes naturally to me. I tally tree species in my log, counting sticks and logs, and estimating crown closure in the canopy. I make a note of saplings coming into the understory, and examine bark for signs of insects or disease. The species of my interest are overhead, not below me. Why should I look down, except to keep from twisting my ankle in a rodent hole?

* * *

It turns out, however, that focusing on the ground can also be useful. Just as birds and their nesting habits tell stories of forest health, so does the understory at my feet. There, amidst old leaves, moss and twigs, another world of green emerges. Ecologists call it part of the native plant community (NPC), which includes everything from moss to old-growth sequoia.

Plants, like humans, tend to grow in communities. They find a neighborhood well suited to their needs and move in. Because small herbs and forbs become established rapidly on a site, they offer obvious clues about the nutrient levels of the soil, as well as its moisture conditions and shade tolerance. Some of them are called "indicator species" because their presence indicates particular soil preferences, moisture regimes, or past disturbance history. Understanding their growth patterns offers valuable information on forest health and history (Kotar, Kovach, and Locey 1988).

Blueberries (*Vaccinium* spp.), poison ivy (*Rhus toxicodendron*), and Labrador tea (*Ledum greenlandicum*) are three examples of forbs that require low levels of nutrients and moisture. Not surprisingly, they thrive in xeric conditions. When I spot these plants in the understory, I know I will not find high-quality hardwoods nearby. Theirs is the low-rent NPC district. Other species, including bloodroot (*Sanguinaria canadensis*), sweet cicely (*Osmorhiza claytoni*), and leatherwood (*Dirca palustris*) require more nutrients, loamy soil, and greater access to moisture. They flourish in the high-rent NPC district where fine hardwoods such as cherry, maple, and walnut can be found. Many plants are not indicator species and will adapt to a variety of conditions. Wild geranium (*Geranium maculatum*), large-leaf aster (*Aster grandiflora*), and beaked hazel (*Corylus cornuta*) are habitat generalists, neighborhood adapters, wanderers. You can find them almost anywhere in the upland northern forests.

Whatever the NPC, these forbs have valuable stories to tell. If your restoration goal is to increase the oak component for wildlife habitat, it is critical to know what species of oak will be better adapted to the local conditions. Blackjack oak prefers dry conditions. Live oak prefers warmer temperatures. Northern red oak prefers moist surroundings. Indicator species and NPC habitat classification offer a powerful diagnostic tool for resource managers.

* * *

And forbs are more than useful markers for field inspections and ecological surveys. They are valuable and beautiful in their own right, and many landowners enjoy their blooms and berries.

On a May morning, spring ephemerals including hepatica (*Hepatica* spp.), trillium (*Trillium* spp.), or marsh marigolds (*Caltha palustris*) will brighten a walk in an upland or wetland forest. Later in the summer, when mosquitoes and ticks add new challenges to fieldwork, brilliant Indian paintbrush (*Castilleja coccinea*) or subtle orchids will surprise and delight even a veteran forester in a cruising vest.

Some plant species are delectable, others medicinal. Morel mushrooms (*Morghella* spp.) offer a delicious spring treat, but don't ask a mushroom hunter to give away their favorite spot. You will need to find one for yourself. Wild leeks (*Allium tricoccum*) can improve a spring salad or stir-fry dish, and American ginseng (*Panax quinquefolia*) is prized for its healing roots and as a medicinal tea. There are more layers to the woods than there are to an onion, and looking down may be as rewarding as looking up.

In almost every season, certain species offer rewards to the careful observer. The berries from blackberries (*Rubus* spp.) or blueberries (*Vaccinium* spp.) may be plentiful in August. In September, sumac (*Rhus* spp.) turns a vibrant red and pods of milkweed (*Asclapias* spp.) open to migratory monarch butterflies. In November, the leaves of bittersweet (*Celastrus scandens*) and winterberries (*Ilex* spp.) glow orange and red, and their berries can be cut and dried. During the winter holiday season, greens pruned from black spruce (*Picea mariana*) or white cedar will brighten entryways and decorate shops. All that is required to enjoy them is a patient eye in the woods and a sharp pair of pruners.

The plants, the birds, the soils, and the water are all elements of an integrated approach to forest management and restoration. As we learn of their roles, we gain knowledge of their relation to the trees above. These interrelationships offer us tools to treat disturbed and degraded systems. In this way our efforts are more likely to succeed and we are more likely to enjoy the process.

CHAPTER IV

FINANCIAL OPPORTUNITIES

Embarking on a Restoration Investment

More than 30 years ago, as I prepared to embark on a forestry career, I went in search of land. I was armed with curiosity and a modest inheritance from my grandmother. At that time, raw land was inexpensive, and forestland was going for less than $200 an acre in northern Wisconsin and Minnesota. An acre of woods was selling at the same price point as the latest technology: a compact disc player. I examined the two investments side by side and asked myself which had better value: 43,560 square feet of land or a small black music box? The choice seemed obvious to me.

A long search for land ensued. Twenty wooded parcels passed under my boots as I combed the northern forest for parcels that would fit my needs and budget. Some properties were large and unaffordable. Others had been cut hard by overzealous loggers. I couldn't see past their degradation and recovery time. Still others were more than a three-hour drive from my urban apartment. I set two hours as the maximum length of time I was willing to drive to reach a property. I wanted to spend more time on the land than in my truck.

There was a 40-acre parcel in northern Wisconsin that I found appealing for many reasons. Primary among them, it contained about 15 acres of white pine seedlings, growing in the shadows of mature aspen and birch. Now those small trees, with the history of white pine on their shoulders, got my blood moving. I saw potential in the seedlings—the potential of helping to restore a cover type that had largely been lost. I knew from experience that white pine was difficult to regenerate on most sites, and had already experienced planting failures of the species from competition and deer browse.

However, there were also drawbacks to this piece of paradise. The road access was poor, and the owner was unwilling to compromise on what I considered a reasonable price, so no deal was struck that summer. Somewhat discouraged, I took a break and went back to the city. When I got there, the first thing I did was buy a new CD player. The music was soothing to my ears.

* * *

Six months later, while listening to my new source of sound, the phone rang. It was the Wisconsin landowner who was now willing to reduce his price for the land. Winter had arrived and he needed cash. Over the phone, he agreed to my original offer with a substantial down payment and a five-year contract for the deed. I hired an attorney to write up the paperwork and went north to close the deal. As I drove into the winter darkness, I felt excited, like an expectant parent. Finally, at an unexpected moment, my search for forestland had borne fruit.

The following spring, when the snow had melted, the young white pines welcomed my boots. In the half-light of a mature aspen canopy, thousands of pine seedlings and saplings greeted their new steward. These trees represented both a past and a future. Their soft green needles spoke of the tradition of the nineteenth century white pine industry in the

Lake States. The conditions for their growth were optimal. The soils were light, and the deer population was low. I was spared the time and expense of site preparation and planting. I was also spared the embarrassment of another regeneration failure. The trees had already risen above the hazel, raspberries, and deer browse line. Here was a patient investment, the start of an enjoyable long-term venture.

Over the next few years, I used my newly acquired forestry skills to take an inventory and tally hundreds of cords of quaking aspen. Because the precise property lines were unknown, I hired a surveyor to put in corners and improved the access with a new road from the north. With the help of an experienced state forester who knew the local markets, we set up a wood sale on fifteen acres of aspen and cleared enough profit to pay for the survey and the final two years on the contract. The land was mine and the white pine seedlings were now over my head.

This was my first experience in purchasing forestland with restoration in mind, and the investment has done well. Over the span of 30 years, the white pine have matured into five-inch saplings and the challenge of blister rust has remained absent. I have also conducted two more wood sales of oak and aspen during strong market conditions, and been rewarded with income three times the original cost of the land.

With some of the profits, I hired a timber stand improvement (TSI) crew to girdle three acres of young maple to release more young pines from heavy shade. And all this occurred as land prices rose and the small pines continued to grow.

* * *

Forestland is not an investment for everyone. For one thing, it is not a liquid asset. For liquidity there are equities, bonds, or even cash or silver. Real estate is less liquid, and raw land is a distant cousin of other assets that are simpler to hold or transfer. And yet, for many landowners and careful investors, raw forestland is a choice investment. For one thing, it is a "hard" asset, less subject to the policies of the Federal Reserve Board or the hedge funds of Wall Street. It is also productive, whether as hardwoods in Appalachia, yellow pine in the south, or Douglas fir in the Pacific Northwest.

If forestland appeals to your tastes, diverse forest investments exist all across the United States, many of which offer the opportunity for conducting restoration efforts. There is hunting land, land with topography, or land with water. There are forests in warm places, in the mountains or near large urban centers. With so many choices at hand, how can one begin to evaluate forestland without getting stuck or going deeply into debt? A useful starting place is a list of reasons *not* to invest in woodland, reasons that have plagued investors and speculators in the past, which can help you avoid many potential pitfalls.

At the top of the list, in my opinion, are two economic approaches that should be seriously avoided: speculation and passive investing (Table 4.1). Speculators are by nature restless, and restlessness is the last thing you want in forestland investments. With forestland, patience, often great patience, is a primary virtue and a key if restoration is one of your goals. Patience is needed to investigate properties and begin to analyze the history of the land. Are hilly terrain, conifers, or hardwoods the best choice for you based on your interests? How recently was the forest degraded, and how severely was it abused? What are the local wood markets?

10 Reasons Not to Invest in Forestland	
1.	You are expecting an immediate return.
2.	You are looking for a passive investment.
3.	You don't enjoy spending time on the land.
4.	You are not interested in your neighbors or the local culture.
5.	You believe that land stewardship is simple and straightforward.
6.	There is no legal access to the property.
7.	You don't know where the property lines are.
8.	You have not done a search for easements or mineral rights.
9.	You think that you can go it alone with managing the property.
10.	Your family is uninterested in your investment.

Table 4.1: 10 reasons not to invest in forestland

Patience is essential while waiting for real estate and timber markets to turn favorable and the price of degraded land to make economic sense to you and your advisors. Don't be afraid to crunch a few numbers and do the math. And most of all, enjoy the search and get to know the land.

A second reason not to invest in forestland is that land, unlike a mutual fund, is not a passive investment. It's not for rocking chair personalities. Land requires a hands-on approach. Be prepared to spend time learning about the trees and plants growing on your property and what conditions they favor or dislike. Be prepared to hop on a tractor to clear brush or to hire contractors to plant, put in roads, or clear trails. Also be prepared to greet your neighbors and listen to their thoughts.

Good neighbors often have deep local knowledge. They may have worked with the local contractors, loggers, and nurseries. They probably know a local mechanic or a supply store. Neighbors also make great watchdogs when you are absent. If you are not a hunter, consider giving them a chance to selectively hunt the land. You might be surprised as to how they will care for it and protect it from outsiders. And, should you ever wish to sell, good neighbors may be keen for the opportunity to expand their holdings.

In short, resolve to spend time on the land and to get your boots dirty. With degraded land, this is especially important. Perhaps the land was grazed or farmed. This is a common phenomenon in the Midwest. Closely examine the soils and what is left before you plunk down earnest money. What species are coming in on the edge of the fields? What is the quality of the remaining woodlands? Have a forester take an inventory. The value of the "growing stock" will affect your investment down the line. For some, return on investment (ROI) is a low priority, but most of us enjoy the benefits of lower costs and additional income. The small cost for an accurate forest inventory will be amply rewarded by a more knowledgeable and profitable purchase. It is likely there will be many parcels to choose from. Pick one that is right for you.

Once you have pulled the trigger and the property is yours, retain your patience. There will be annual taxes to pay and years when the greatest return will be to watch the trees grow. If you are the speculative type, you will be restless and nervous and should refrain from forestland investment. It is for turtles, not hares.

However, if you enjoy your time on the land, whether for hiking, hunting, or chopping wood, you will weather the setbacks. There is pleasure in burning a brush pile in autumn and in tapping a maple tree is in the spring. The pleasure may be as simple as the smile on your face after a day in the woods.

* * *

It is true that, at first, an investment in forestland will eat up capital and time during the course of its purchase, its maintenance, and its initial restoration steps. But if you appreciate the land, if you love to walk it and to work on it, you will be rewarded many times over. There is more than economics and statistics to owning and caring for forestland. There is clean air and solitude. There is the sound of the wind in the canopy. There are morels, spring ephemerals, and young pines rising from the shadows. There is silence in winter and a chorus of frogs in the spring. This long-term commitment also offers an opportunity to restore your spirit from some of the stresses of modern living. The reward cannot and should not be measured solely in dollars and cents.

A Savings Account in Oak

When I was young, my father impressed upon me the importance of a savings account. At seven years old, proudly clutching my $10 in savings from weekly allowances, we rode down to the local bank and opened a new savings account in my name. This was in the days before online banking. I received a small green passbook with the account balance written in ink in the right-hand column. When I returned home with my father, I felt as if I had completed one of the rites of passage into adulthood. Carefully, I filed the passbook away in a small drawer of my desk. Each time I removed it, I was reminded that saving for the future was one of my father's maxims of life. Perhaps that philosophy could help me in my career as a forester?

* * *

With the purchase of the white pine site safely under my belt, I realized that degraded or disturbed sites held potential for long-term investment and restoration. On these lands, I realized a potential that others generally overlooked. At that time I was employed as a researcher on an oak regeneration project for the local university. Our focus was on how to improve the natural regeneration of red oak on quality hardwood sites. It turns out that oak is easy to regenerate on dry sites, but very difficult to bring back on mesic sites where its value for wood products is high. Because of these silvicultural challenges, it was a fascinating project that made me think about the potential of degraded oak sites. In the research project, I inventoried more than 100 oak sites on public lands, most of them "high-graded" from past harvesting practices. I also had a chance to work with two premier oak researchers in the country, Paul Johnson and Rod Jacobs. They helped me to learn how red oak was adapted for disturbance and how it responded. This got my juices flowing. In my spare time, I went in search of a new investment: one in red oak.

One 120-acre site in Wisconsin impressed me immediately. Loggers had recently hit this parcel, and they had hit it hard. In forestry circles, we often talk about high grading and *diameter limit cutting*. *High grading* is a technical term for taking the best and leaving the rest. *Diameter limit cutting* is related to high grading, but is perhaps more obsequious. Under this method, all marketable trees over a predetermined diameter are harvested. Properly utilized, the diameter limit is set high and only a few mature trees are harvested. More often in practice, the diameter limit is set low and the forest is high graded of all profitable species. In the case of this parcel, both had occurred. The land had been stripped of all the valuable red oak over 12 inches in diameter. Now that the high value was gone, the loggers had put the stripped land on the market at a low price.

Why was I interested in such a piece of land? It looked rough on first glance, with limbs, stumps, and debris still littering the site. The remaining oaks were small or had serious defects. There was also a large gully down the main logging trail that posed access difficulties. But as an oak researcher, I could see past the present damage. The soils on the site were loamy. A quick check of my NPC charts and forbs verified that high-quality hardwoods thrived on such soils. I made a cruise consisting of 30 random sample plots well spaced on the land. With these plots, I conducted statistical extrapolations back in the office.

My cruise offered a wealth of information. It provided me with a list of all the marketable species, and there were some surprises. Saw logs of black cherry (*Prunus serotina*) and white oak (*Quercus alba*) showed up along with hundreds of cords of northern red oak in the 10-inch and 12-inch diameter classes. These trees had straight boles and their crowns had been released by the recent cutting. My cruise also measured tree ring growth and showed excellent growth patterns on the remaining red and white oaks.

I also sampled the understory. There was sapling sugar maple and basswood, but no buckthorn (*Rhamnus cathartica*) or prickly ash (*Zanthoxylem* spp.), two invasives that make oak regeneration difficult. And the maidenhair ferns confirmed that the site was mesic and well suited to supporting quality hardwoods. Here was a more complete glimpse into the health of the forest.

I thought of my father's dictum about savings accounts. The straight oak trees would accrue interest every year. As their volume and quality increased, this forest would mature and add to baseline capital. And so I made a reasonable offer and it was accepted. I had just become the owner of another parcel with restoration as its goal.

* * *

It has now been more than 25 years since I purchased the property. In the first 10 years I repaired the main trail and let the land rest. I also got to know my some of my neighbors, who were hunters and had adjoining land on the back 40. Land prices began to climb and my neighbors were interested in connecting their land and improving their deer habitat. They also knew that the oaks had value, both as wildlife mast and as future saw logs.

I had a more accurate survey done on the back 40 and I struck a deal with my neighbors. When the sale closed, I pocketed more than I had paid for my initial investment for the entire 120-acre parcel, and I still had 80 acres of oak woodland that was increasing in value and testing my restoration skills. My father's dictum about a savings account was at work.

After owning the land for 15 years, it was now time to think about the oaks and begin some active restoration management. You may be asking yourself how this approach is related to restoration. It appears to be merely a patient approach to good forest management. My response is simple. In the first place, the land had been disturbed and degraded. This is the first principle of restoration. Degraded forestland comes in many shapes and colors, but not all of it is mining spoils or pastureland. With Mother Nature's help, it was now time to make an active effort to restore the health of a veteran oak forest. This is the second premise of restoration: an active human hand with a plan and a method to measure the results is needed to reach the stated goals. But what was I restoring?

Let's take another look at forest terminology and how it affects our efforts. In the United Kingdom, they do not use the term *old growth* as we do in the United States. They prefer to classify their forests as working forests, veteran forests, or ancient forests (Watkins 1990). Ancient forests are the oldest ones in British terminology, and comprise those that are more than 400 years of age. Veteran forests are those in late maturity, usually 200 to 400 years old, depending upon the species. It turns out that around the world, veteran forests are in short supply.

Most of the veteran forests have been cut in the past 200 years for the products they provide. In a real sense, they are threatened.

My goal on my 80 acres of oak hillside was to restore a veteran hardwood forest that was at risk of being lost to sugar maple and logging. This goal goes against the grain in some forestry circles. For some managers, the goal is to maximize fiber production over a relatively short cycle. For northern red oak (*Quercus rubra*), the standard recommended age to final harvest is 80 to 120 years. In my experience, this is short-term thinking, particularly because red oak is a difficult species to regenerate on good sites and often grows well for more than 150 years. Restoring it to a veteran oak forest will also provide more than merely economic benefits. It improves the groundwater aquifer and wildlife habitat for deer and many other species. As the owner, I had the opportunity to think of the oaks the way many European managers do: in terms of their more-natural, longer lifespans.

* * *

Now that I had a goal of cultivating a veteran oak forest, where was I to begin, and who could help me? It is one thing to have a restoration goal, but another to implement it. After 15 years of rest, my red oak hillside was in need of an intermediate stand treatment (IST). An IST is a silvicultural tool foresters use to improve the density, spacing, and health of a stand of trees (Smith 1986). My IST was also designed to remove many of the small maples that were impeding oak regeneration and to give the red oak "crop" trees better growing conditions. One way to think of this is as a thinning in your garden to improve its growth and vigor. In addition to providing modest income, these ISTs improve the growing conditions of the forest. The remaining trees have more light to expand their crowns and more moisture for their roots. Young oak seedlings also have better survival rates when they have less shade over their heads. The result is a more resilient forest that is less prone to drought and disease (Fig. 4.1).

Figure 4.1: Pole timber red oak that have been thinned.

* * *

Because my home turf was 150 miles to the west, I hired a local forester to mark the trees to my specifications and to set up a wood sale. This experienced manager knew his local markets and the reliable loggers. It made sense to bring him on board to guide the process because timber sales are complex and can involve many variables that require evaluation. The boundaries of the sale and the property lines needed to be well established and carefully flagged. The thinned trees were each marked with paint to guide the loggers, indicating which trees were to be removed. It takes an experienced eye and a careful hand to properly mark a hardwood stand for an improvement thinning. Residual tree spacing, canopy closure, and space for machinery to operate all must be taken into consideration. An accurate cruise, a good prospectus, and a mailing list of carefully selected loggers are also essential. When the bid opening for the logging operation occurs, the forester will track the bids, whether in an open or sealed auction. And this is just the beginning of the process. Where should the log landing go? Will new roads be necessary? Are there stream crossings? Is the sale expected to take place during a winter or summer?

Wood sales do not happen overnight. When a logging contract is signed, the operator is usually given extended time to complete the sale; a couple of years is common. A down payment is made and then there is a lull; a quiet pause before machinery moves onto the site. This is a good time to review your restoration goals. Will you need to follow up the postharvest with additional site preparation or planting? If so, what is the timeline for the next step? What will happen to the slash piles of debris from the sale? Who is responsible for damage to trails or to residual trees should it occur? What happens if the weather turns wet or markets change? Talk over all of your concerns with your forester, colleagues, and friends before signing any contracts or making any commitments.

* * *

The most critical time of a harvesting operation is the startup. This is when potential problems are easily identified and best solved. Is the landing large enough for semis and wood processors? Are your neighbors aware of your project? Are the boundary lines easy to spot in summer foliage? How are scale slips to be handled? Although a professional should handle the essentials, owners can make their presence known without becoming overbearing. Don a hardhat when visiting an active site. Don't be afraid to ask questions, but acknowledge that the logger needs to work to make a living.

An active wood sale is a dramatic, energy-filled environment. If you have not observed high-tech machinery in the woods, prepare for a moment of revelation. Stand back and enjoy the show. The operators know what they're doing, and they move heavy material with surprising ease. Tracked vehicles swivel into place and clamp a tree. The buzz of a saw occurs from the safety of a cab. An aspen shivers, separates, and drops onto the duff. Roller blades spin and grind off branches. Another buzz and the tree drops into sticks. Three sticks, four sticks, five sticks. They appear in the blink of an eye. The processor turns, repeats the process, and then moves on. As the day lengthens, the woodpiles grow. A forwarder snakes its way through the clearing and gathers logs into its belly. A calm returns to the site. The woods appear more open and light reaches the forest floor. You may have never seen these woods bathed in so much light.

After a few days, or perhaps a few weeks, the processor's task is almost complete. On the landing, the woodpile continues to grow. Semi-trailers appear and the woodpile starts to shrink. With each load, 10 or 12 cords move off to the mill. The forwarder comes out of the woods with its last load of logs. It is time to load it on a flatbed for the next site.

This is the time to tidy up the site. Have the trails been bladed to preharvest condition? How much slash is left on the landing? Can it be chipped, burned, or buried? Are there damaged standing trees near the trails? Should those be felled or left in the woods? Are all the load slips accounted for? After all, you want to be paid for all of the wood, not some of it.

Step back and breathe a sigh of relief. The harvest has gone well. The forest is disturbed and disheveled, but it will not stay that way for long. The new haircut will take time to appreciate. The net result of this sale will be an increase in your banking account and a healthier forest, bringing you and your forestland one step closer to your restoration goals for a veteran forest.

The Costs of Restoration

"For 10,000 years oak was the prime resource of what was to become the western world."
—*W.B. Logan*

Not all restoration projects will put money into your pocket. Most, in fact, will require capital expenses that need to be thought out with care. Many well-intentioned forest restoration ideas fail to be implemented and get stuck. Stuck from a lack of planning. Stuck from opposition from interest groups and neighbors. Stuck from bureaucracy. But mostly they get stuck because of money.

Restoration has a reputation as an expensive enterprise, particularly in the public sector, because the costs of design, implementation, and monitoring often run into four figures per acre. I recall giving a talk on restoration at a national conference. During my presentation, I bemoaned an example in the Southwest where the costs of Ponderosa pine restoration on federal land ran more than $1,500 an acre. After my talk, a manager from the Southeast came up to me and informed me that some of his restoration projects ran two or three times more on a per-acre basis. He then informed me that he managed a reserve for the armed forces, and was charged with longleaf pine restoration on the base. I must have looked surprised by his words, but they got me to thinking. On how many acres can private landowners or managers afford to practice restoration with costs in this range? The answer is self-evident: usually not many. Of course, there are good reasons why large sums may be needed to get the ball rolling or to set examples for others to emulate. But we need to find cost-effective approaches to restoration rather than spending our time justifying big bucks and expecting others to follow. Otherwise, most private landowners will be eliminated and our reputations may be tarnished.

How do we accomplish this? My job as a forestry consultant is often to assist landowners in understanding the potential of their land, potential they may not be aware of. For example, there are many thousands of acres of degraded oak savanna in the Midwest and today oak savannas are more threatened as ecosystems than are white pines. From experience I can often spot these degraded savannas from the highway: the topography, the bur oak sentinels, and the understory invasive species usually give them away.

Because many of my clients are not aware that they may have a degraded old oak savanna on their land, the first step is to let them know. This can be done on a walk, in a management plan, in a newsletter, or at a workshop. When made aware of a unique feature, they will often respond like my neighbor with his bald eagle nest. Their eyes will light up (Fig. 4.2). The first step toward restoration has been taken and it did not cost anything to take it.

Figure 4.2: Bur oak savannah

* * *

After the spark has been lit, the next step is usually in the owner's hands. This is where economics comes in. How can a restoration project be undertaken within a limited budget? Sometimes the answer is tied to income-producing projects that can piggyback on the restoration effort. A pine plantation may be thinned and white pine planted with part of the proceeds. Mature aspen from the back 40 may be harvested and part of the income used to remove invasive shrubs or restore an old agricultural field to native prairie. The local conservation district may have a cost-share program for improving wetland habitat or restoring native species. And these are just three examples from projects I have completed in which income from the land helped landowners to restore their degraded woodlots.

Sometimes restoration projects depend more upon the landowner's skill set than their pocketbook. Many owners have access to tractors, brush saws, discs, and plows. Once they are aware of the potential to restore part of their land, they may be motivated to clear out some buckthorn, to plant native oaks, or to create a firebreak. They may even have friends who work for the local volunteer fire department who will assist in designing and implementing a prescribed burn. The required capital is primarily in time and labor.

In other instances, there are natural partners in restoration and costs can be shared. A conservation organization may have funds to help a habitat project for one of its favorite species; the local soil and water district (SWCD) may provide planting stock at a low cost; the highway department may wish to help maintain a road right of way for native plants to flourish. All of these cooperative options can be explored or implemented to reduce out-of-pocket expenses and jumpstart projects.

Restoration usually does not take place overnight. It takes time to understand the causes, collect timely data and design a careful plan. Often it is critical to scale back ambitious projects to bring them within budget constrictions. Just remember the most important rule of success in Hollywood: perseverance. It is not the size of the project that is critical. Rather, it is getting it funded and on the ground. There will be a learning curve. There will be monitoring of the results. What maintenance is required? What planting stock will be preferred? With all of this knowledge, you place yourself in a better position to make decisions that affect all of the land you care for. And you don't need to break the bank to do it.

Bark and Boughs

The majority of my clients believe that the economic value in their woods derives solely from timber products. While they admit that there may be a few wild edibles growing here and there, who wants to take the time to pick them or take the risk of getting sick? Surprisingly, it turns out that profits from the woods can come from many sources. The experience of one of my clients, an older gentleman called Mr. E., is a case in point. Mr. E. owned 120 acres of raw land that had been in his family for more than 60 years. When his family bought the land, it was an early twentieth-century farmstead with cattle and agricultural fields for hay and crops. But the winters were too hard for farming and the soils were better suited to trees than corn.

Early in his adulthood, Mr. E. began to restore his old farmstead fields to pine and spruce forestland. He joined the statewide tree farm association and became a founding member of the local forestry cooperative. He took the cattle off the land, improved his degraded woods, hosted field tours, and brought in experts to share the latest information on forest management techniques. He always had a smile on his face and a kind word on his lips. In short, he became a model of forest stewardship in his neighborhood and many of his efforts were focused on restoration of degraded land.

Mr. E. called me one morning to ask for a new management plan. His old plan was outdated and he liked my approach to forest management and restoration. After all, he'd been practicing restoration since long before I had field boots. I reviewed his old plan and made notes on the history of the land. Then, with recent aerial images in hand, I drove out to his place and we began a ritual walk. We discussed the history of the farmstead and talked about thinning his pines and improving the trail system. We stood by a couple of large diseased butternut trees (*Juglans cinerea*) and I gave him the contact information of a local wood-turner who was looking for butternut logs. In our walk that morning, we covered many topics, including whether to restore his last agricultural field to trees or prairie, how to clean up from a recent aspen harvest, and where to plant more spruce in the understory for wildlife cover and diversity.

But there was one area that we did not discuss in detail: a sizable lowland conifer stand that grew on the northern side of the property. Because it was a bog and difficult to traverse, he rarely visited the area, and I made only a cursory inspection and notes that day. The land was dominated by black spruce (*Picea mariana*) and tamarack (*Larix laricina*), and had a high water table. Most of the trees were stunted, growing poorly, and of negligible commercial value. Or so I thought.

Both Mr. E. and I assumed that this part of his forest was good for reclusive wildlife such as bobcats, mink, wolves, and owls. It was also an important source of clean water for the area because the wetland soils filtered and held sediments and nutrients. The quality of the groundwater and nearby lakes was improved by this large natural filtration system. We also assumed that the small, slowly growing conifers had little potential as a source of income for the farmstead. But changing market conditions would prove that with regard to this last assumption, we were seriously mistaken.

Time passed. With a new management plan in place, Mr. E. enrolled his last field in the conservation reserve program (CRP) and planted it with a mixture of shrubs, hardwoods, and conifers. He thinned his red pines and put in a new trail. Then, one fall morning, he got a telephone call from a local couple with a small Christmas "greens" operation. They were looking to harvest some spruce tops in late autumn to ship down to city markets. Could they stop over and harvest an acre or two in the bog? It would only take a day, and they would pay him for the greens at the landing. (Fig. 4.3)

Intrigued by the unexpected offer, Mr. E. agreed to the experiment. Then he got on the phone and started to ask questions of his forester about this "bough" business. Was there much of a market for boughs? How did the harvesting take place? Did the harvest damage the wetlands in any way? Because this was new territory for me, I referred him to a county forester who had worked with spruce-top management. The learning curve began for both Mr. E. and myself.

A few weeks later, the "bough" couple showed up with a high flotation four-wheeler trailered to their truck. Mr. E. made certain to be there when the operation began. They used the ATV for access to the wetlands, and only removed the top 2 to 4 feet of the stunted spruces. When they finished that day, they left the wetland in good shape and paid Mr. E. on the spot. The check was larger than he expected. In a parting gesture they asked if they could return the following year and harvest in a wider area of his lowland conifers.

More than 10 years have passed and Mr. E. shares his story with a wide grin. His is a grin that comes from unexpected success. The "bough" couple grew their business and returned three times to his underappreciated spruce bog. He shares pictures of the spruce swamp with his audience. It looks almost untouched, with healthy trees sprouting new leaders, and no ruts in the wetland grasses. And there are more photographs of the boughs piled up in his field, awaiting a refrigerated

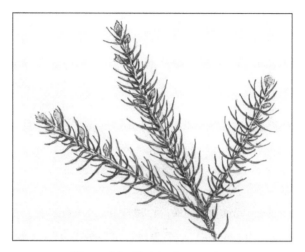

Figure 4.3: Sketch of spruce tops

trailer. Off to the side stands a proud owner. He has cashed checks well into four figures for his swamp spruce, checks that have more than funded his taxes, wildlife plantings, and a new tractor. Clearly, he has reason to smile.

I am reminded that Mr. E.'s spruce bog was part of the woods that both the landowner and his resource manager had neglected as a priority management area and source of income. And yet, the spruce bog had earned Mr. E. more income than on all the rest of his upland forest acreage combined. Perhaps there was a lesson there.

I, like most resource managers, was not immune from preconceived notions about woodland values. In this regard, foresters are no different from other professionals: real estate agents, accountants, and financial planners, to name a few. The appraisals we make are based on history, past markets, past growth, and past profits. Reasonable assumptions are made and projections are modeled for the future, but these projections may be inaccurate or contain outdated assumptions.

A hundred years ago the primary forest market in the Lake States was for large diameter pine. Aspen (*Populus* spp.) was considered a weed tree with almost no commercial value. Then, following World War II, the markets changed dramatically. The pine-fir building market moved west. On the cutover Midwestern forests, second growth aspen grew rapidly and soon became an inexpensive raw material for a new breed of pulp and oriented strand-board mills. In the late twentieth century, with a building boom to feed it, the price of aspen began to rise. Before long, forest landowners realized that they could sell aspen stumpage for more than the original cost of their land. What had once been a low-value species had become a high-value fiber.

More recently, markets have shifted again. Paper mills have declined in importance as the country's use of paper products has dwindled. With the shuttering of plants and the majority of the mature aspen already harvested, *Populus* spp. is no longer a hot commodity in the northern woods. Markets for other products, including spruce boughs, have appeared. Where the market will head next? Carbon credit storage? Birch boughs? Maple syrup? Basswood carving blocks? High-value hardwood veneer? Or some unknown product for which the value has been ignored? While I do not have a crystal ball, clearly the forest holds more than saw logs and pulpwood sticks for enterprising entrepreneurs.

Spruce boughs are a fine example of an unexpected market niche. Maple syrup is another that is on the increase. At least a dozen of my clients regularly tap their sugar maples in late winter and early spring. Some of these operations are small, numbering perhaps 50 taps. They produce holiday presents for fabulous pancake syrup year round. Other woodland owners are more ambitious and tap up to 1,000 trees per year. Tapping 1,000 trees takes a hearty set of lungs and hundreds of buckets and spikes. The sap is heavy and the snow is deep in March. The work, however, is satisfying to these woodland owners. After a long winter, it is the earliest outdoor season. As the sap steams into syrup, the producers enjoy aerobic exercise and may earn significant income from their labor.

In addition, some of them use forest thinnings from their own woods to heat their stoves. As "syruping" has grown in popularity, more private landowners are active in managing their maples and veteran sugar maple forests are on the rise.

There are other markets on the horizon as well. Forest owners in California and some European countries can already earn credits from sequestered carbon: carbon that is stored by leaving old growth or veteran forests alone. Carbon credits serve as a new market opportunity, and these credits may be traded on the open market and purchased by utilities companies and heavy manufacturers to offset their carbon burn. Who would have prognosticated such a market 50 years ago?

Niche markets also exist for many additional plants and trees, including birch bark (*Betula papyrifera*), black ash burls (*Fraxinus nigra*), ginseng (*Panax quinquefolia*), and shitake mushrooms (*Lenticula* spp.). Perhaps you have one of these species in your forest, and perhaps they can help you fund a restoration project down the line. The next time you walk in the woods, be open to the unexpected values there. Like Mr. E., you too may end up wearing a smile of success.

Learning the Financial Ropes

Most restoration projects take time to design and capital to implement. Prescribed burning to restore native prairies, oak savannas, or pine habitat requires a trained crew and careful preparation. Removing well-established invasive species is time consuming and labor intensive. Planting or seeding degraded sites is more complex than hoedading seedlings into the ground. There is site preparation, a skilled planting crew, and browse protection all to be considered. As a result, understanding the complete economic picture of restoration efforts is essential to long-term success.

The final component of the economic leg is, in fact, found far from the woods. It is found in the world of accountants, assessments, taxes, depreciation, and cost sharing. Although these issues don't offer the drama of a prescribed burn or the beauty of coneflowers in the summer, they are critical to most successful restoration efforts. It behooves family forest owners, managers, and those considering land investments to spend time learning the complex financial ropes of restoration and ownership.

Let's begin with taxes, our best friend and worst enemy. Taxes are an essential part of our free enterprise system. They maintain roads and build schools. They fund fire departments and make life easier in retirement. At the same time, as taxes rise, they can be a cause for distress and can lead to sleepless nights. Property taxes are an important subset in the world of taxes and a major source of revenue for local governments. In each municipality, the assessor's office carries a wealth of information about raw land and its assessed value. This is a valuable place to begin research on local tax codes. In most states there is a variety of tax rates depending upon whether the land is homesteaded, farmed, or owned for recreational or investment purposes. If you are a private landowner, do not hesitate to speak with your assessor to understand how these classifications are determined and what options are available to reduce the tax burden. How is the local assessment calculated? Does it accurately reflect the current condition of the land? Are deductions available for easements, wetlands, or active forest management? In many states landowners who actively manage their woodlands, with written stewardship plans to guide them, qualify for a significant property tax reduction. Although the guidelines vary from state to state, these deductions or rebates can make a difference in the long run between profit and loss in the woods; it may mean the difference between keeping the land in the family or parceling it to a developer.

The rationale for government tax programs that benefit forestland owners is that healthy forestlands benefit everyone in the community. Healthy and thriving forests produce cleaner water for aquifers. They remove carbon dioxide from the atmosphere and replace it with oxygen. They support local business and industry and provide diverse habitat for wildlife. It makes sense that landowners who care for their woodlands and are considering restoration projects should be eligible for reduced tax rates, yet many forestland owners are not aware of these programs. A trip to the local tax assessor may sound dull or confusing, but it may bear significant financial rewards. Once again, I am reminded of my father's dictum on savings accounts.

* * *

Learning the financial ropes also entails understanding the growth patterns of forest species; tracking the costs of capital improvements; and calculating whether your land qualifies as an investment, a business or a recreational parcel. This analysis requires specialized expertise, and a team approach is warranted. A forester can establish the cost "basis" for the trees at the time of purchase. A tax accountant can establish the depletion allowance for your next wood sale. An attorney can assist with easements, property rights, and perhaps establishing a family trust.

Because the purchase of raw land is actually the purchase of a bundle of rights, it is critical to understand these rights and the responsibilities that go with them. For example, the deed may convey mineral rights, partial mineral rights, or merely surface rights to the new owner. In some parts of the country, these mineral rights are far more valuable than the surface rights. Determining who owns the mineral rights on the back 40 (the state, a mining company, or the prior owner, for instance) is important. Fortunes have been made or lost based on this ownership information. Your attorney may take the research one step further and explore whether any legal easements exist on the property. Are there rights for township roads, power lines, or a neighbor's access? Understanding these rights will affect how you manage the land.

Although expert advice may at first seem unnecessary or expensive, in the long run, a professional team will save you headaches and financial mistakes. If you choose your team carefully, each member will bring their experience and knowledge to the table. In this manner, you become the captain of a safer, more seaworthy financial ship.

Above all, do not despair as the complexity of your land investment increases. Learning the financial ropes requires time and patience. Although it may be less memorable than listening to a loon or tracking a deer, it is the side of the equilateral triangle that can make or break your tenure with the land, and it can be done in the wintertime when the snow is deep, or in the heat of a summer night, when the bugs are biting. There is a learning curve to all endeavors, and financial responsibility has many rewards. To heed it gives you a leg up in your process of successful forest stewardship and restoration.

CHAPTER V

CULTURAL CONNECTIONS

Restoration in a Mining Town

I lived in the old mining town of Crosby, Minnesota for 15 years. My forestry office was on the second floor of an early twentieth-century bank building on Main Street. The first floor of the old brick building had been restored to a comfortable air-conditioned real estate office. Most of the buildings in town had not been as fortunate.

Crosby is a town with a rough history. Many of the old shops were closed when I moved to town. The downtown was gloomy; doors were boarded and "for rent" signs greeted visitors and locals alike. These were all the signals of an old industrial town where industry had moved on. In Crosby, the industry had been iron mining.

The remains of the old mines rested outside of town. Many of them were underground. A few were "open pits," visible from the surface. When I first visited Crosby 50 years ago, one of the open pits was still active. It was impressive to gaze into the deep rocky crater from the highway. Inside the mile-wide raw opening, oversized mining trucks kicked up red dust. As the trucks inched down the slopes, they began to shrink. By the time they reached the foot of the mine, these oversized vehicles were no bigger than a child's toy truck in a sandbox. I was always struck how the depth and size of the pit could transform massive earthmovers into toys.

The pits themselves were far from child sized. Carved into gravel and rock, they were strictly a serious work place. The work focused entirely on ore extraction. The early Crosby mines were underground labyrinths with their shafts out of sight. But they were dangerous, and when one of them burst in 1924 and filled with underground water, 41 miners were killed. Not long after the Milford Mining disaster, most of the mining shafts closed and open-pit mining became the norm on the Cuyuna Range.

Open-pit mining is kinder on people but not on the landscape. It begins when bulldozers level the forest and strip away the topsoil. From there, the "overburden" of glacial till is removed. Seams of red ore are exposed and the blasting begins to break up the hard rock. Finally, after all of these challenging preparations, the ore is loaded on small ore cars and hauled to the surface to be scrubbed and shipped.

This process of digging, dozing, and sorting creates mountains of debris and mining "spoils" for open-pit mines. These spoils need to be disposed, and the most efficient method is merely to pile them adjacent to the mines. As the mines grow deeper, the mining tailings by their side grow proportionately. For a mine that remains open for 40–50 years, this creates a literal mountain of tailings. Over the years, the level landscape around Crosby was transformed into a range of hills and holes; it became known as the Cuyuna Range, a veritable mosaic of crushed rock and dust.

* * *

Between World War I and World War II, the mines flourished and this raw landscape turned red. Crosby's iron was high in manganese; an essential hardening ingredient for steel, and the ore was in great demand. The country was hungry for steel: the basic building block for new buildings, bridges, and the weapons of war. But by the late 1950s, the costs of mining in Crosby grew, and the rich mining seams were nearly exhausted. New mines in Australia, Africa, and South America were feeding global demand and the old iron mines of the Cuyuna Range were literally left in the dust. One by one, the mines in Cuyuna Range closed and the town wept. The rail lines to the mines also emptied. Except for an occasional rusted ore car, its sides burning with the color of orange chalk, only grasses and sumac filled the tracks. In town, vacancies became the norm. Many mining families moved on. Others stayed and settled into a survival mode. Except for during the summer, when tourists enjoyed the nearby lakes, the town hardened or slept.

In the open pits themselves, however, another chapter in the story began to unfold. Mother Nature went to work. She began with her most plentiful resource, the region's groundwater. Slowly the barren hollowed-out mines began to fill. At first it looked like a mistake. Had the pumps stopped working? But there were no shovels in sight and the pumps had been removed. From summer to summer, the water inched higher and the old mines began to look like half-filled lakes with rocky slopes. The townspeople took visual note, but this time humans were the bystanders, not the activators.

As the waters rose, the talus and tailing slopes of orange and charcoal mining spoils began to turn green. Quaking aspen (*Populus tremuloides*) seeds blew in with the wind. As a pioneer species, adapted to populate in sand and gravel, aspen is one of the few species that can tolerate the mineralogy and acidity of old taconite tailings. Grasses and small shrubs joined the aspen and the tailing hills took on the look of a young forest. Intrepid hikers began to explore the new landscape. Humans reentered the landscape.

Then, lake trout were introduced to the pits. The lake trout did well in the deep cold water of the old mines and fishermen were not far behind. They lobbied for boat access to the old mines and for parking for their trailers. Even mountain bikers discovered the steep talus slopes and the challenges of the new topography.

All of this took time. Seasons passed when efforts slowed and Crosby languished. There were skeptics and a lack of funding to build new trails; after all, this was an old mining town and slow to shed its skin. But some of the area's residents persevered with a new dream. "Build it, and they will come," they echoed. More mountain bike trails were built. Bike shops opened. Kayaks appeared. Word of a new recreational landscape began to spread.

* * *

I recall first taking my wife to the top of the trail overlooking the old Pennington mine west of Crosby. It was her first trip to the old mines surrounding the town. She seemed surprised at my exuberance as we climbed the hill and I explained the history of the intrepid aspen forest alongside the red rock road. It was a splendid autumn day. At the top of the hill, she caught sight of a wonderland of blue, green, and gold. Here was mountainous topography in the midst of a prairie state. The late aspen shadows reflected in the water from the Pennington mine. The yellow leaves radiated on the hills surrounding us, those human made rock piles built on mining debris. As a cool, clear wind blew in from the north, she casually remarked that all of it reminded her of her family mountain home in West Virginia. I was surprised by her observation, but I shouldn't have been. Restoration of this landscape was on full display, and she never knew the blue lakes beneath her eyes as old mining pits. To her, they were another set of cobalt blue Minnesota lakes.

Who was responsible for this massive restoration project? Humans could only take part of the credit. Mother Nature had done the heavy lifting and opened the door for the area to be reborn as a recreational park.

It has now been almost 50 years since the last mine closed on the Cuyuna Range and Crosby has begun a restoration project of its own. Today, hundreds of bikers seek the area to enjoy the mountainous biking trails of the Cuyuna Range, and the town of Crosby is once again thriving with new restaurants, breweries, and other businesses that cater to the younger generation. The "for rent" signs have largely disappeared.

The Role of Cultural Restoration

Crosby's history shares the power of Mother Nature when it conducts a restoration effort. But it also shares the cultural power of humans to reinvent their habitation. How many ancient cultures have been buried and rebuilt in Europe and the Middle East? A trip to Rome will open one's eyes to a modern vibrant city built over, around and amongst the walls and temples of a 2,000-year-old civilization. If this is not restoration, what is? Culture plays a central role in forest restoration as well because cultural activities and attitudes often shape forest efforts on the land and in the woods.

Recreation is a prime example of the power of restoration. For some, recreation means a summer retreat on a shaded porch in the heat of July. For others, action is imperative. All-terrain vehicles, mountain bikes, and jet-skis capture their attention and bring them in contact with the forest. Eyes light up when speaking of a special hiking trail, a new deer stand, or a recently discovered hawk's nest. You may sit motionless in the autumn, awaiting your prey. I may step softly in the summer in search of an ephemeral lady's-slipper (*Cypripedium* spp.). We may gather family and friends around a grill on a warm spring evening and listen to the chorus of spring peepers. These moments touch the core of family values and are a primary reason to own and protect forestland. They also represent a "time out" from the hectic pace of modern life. Spirits are rejuvenated, batteries recharged. The concept of personal restoration is instilled in the bones of those who spend time in the verdant forest.

* * *

Cultural connections tie us to the land. It is not enough to understand only the ecology and economics of a site. Natural resource scientists and researchers often believe that we have all the tools to rehabilitate or restore landscapes, but all too often in our planning we neglect the people who live there, the cultural community. At our own peril, we ignore the third leg on which the restoration stool stands.

The Chicago Restoration Controversy in the late 1990s bears out this difficulty (Gobster and Hull 2000). The Chicago metropolitan area is blessed with a large public forested reserve system. In Cook County and other surrounding counties, forest preserves have a long history and are beloved spaces for many urbanites. When the news broke that resource managers were going to tear up thousands of acres of these preserves to restore prairie grasslands, the public was shocked. How could this be progress, they asked? Northern Illinois had a geological history of oak savannas and prairies. Before European settlement, prairies were common, aided by tribal customs that included burning. But now there were forests in place of the prairies, forests that most urban dwellers loved. The assumption made by the professionals was that they were privileged to make this change to the landscape. They knew best. But did they?

As resource managers, if we hope to succeed with large-scale restoration projects, we need to keep our audience firmly in mind. This does not mean we merely need to educate the audience, as is commonly assumed. We need, instead, to bring them into the loop, listen to their concerns, and learn their history with the land. Otherwise, despite all of our best advances with science, we will fail to move the needle.

The Chicago area managers found this out the hard way. They ended up at countless county commissioner meetings where they were outnumbered. The press turned against them. Years dragged by in courts and in confrontation. The scientific researchers and managers dream project was put in moratorium by the public process of democracy. After thousands of hours of preparation and research, the net result was an impasse. The county forestlands would not be converted to prairies.

* * *

A few years ago I was invited on a field tour with a group of college students. These students were studying forestry and ecology at a nearby college and their professor wanted to give them a first-hand look at a real-life restoration effort. At the offering of the county land commissioner, we were all taken to a black ash forest in northern Minnesota for a tour. When we reached the site, we were surrounded by swampland and black ash (*Fraxinus nigra*). Over 90% of the overstory was in ash, due to the fact that it is one of the few species to survive in the hydric soils of a peat wetland. The ash trees were in full leaf out, and the understory was sparse. Most of us wore knee-high rubber boots. Some of the students made the mistake of wearing sneakers and learned one of their first lessons in fieldwork. Be prepared for adverse conditions.

We were all there for the purpose of a real life experience and education. For many of the students, it was their first time in a black ash swamp. Although there are more than a million acres of mature black ash stands in Minnesota alone, they are at risk from the latest invasive critter from overseas, the emerald ash borer (EAB). Native to eastern Asia, the insect was brought over on crating and has found a new home in the United States. Because the EAB does not have natural enemies in America, the species has been able to thrive. As a result, it has already destroyed millions of ash trees in the Midwest. Most of these are boulevard trees, but the bug loves ash wetlands as well. The species is spreading north, where it may find a feast of its favorite food and wipe out thousands of acres of ash.

If this isn't a grim enough picture, there is an even darker side to the story. On wetland sites that have been attacked by the beetle in Michigan, the forest has not come back. After the ash canopy trees die, water transpiration is curtailed and the water table rises significantly. The forest is literally swamped and often becomes a reed canary grass and cattail wetland. In Minnesota, the risk is the loss of thousands of acres of current forestland. That would amount to degradation of an important hydrological and ecological resource.

The college students were brought to this isolated site to see an ecological challenge first hand. The foresters and managers in Aitkin County were trying to get ahead of the curve. They were experimenting with partial-harvest treatments in the ash to try to regenerate other tree species before the EAB arrived. They were proactively managing to restore the site to trees *before* it became degraded. It was a race against time. The public managers knew that when the iridescent insect arrived, it would be too late to reforest the site. Based on the EAB's track record, it was likely there would be no forest at all but rather a wetland.

Hence, the purpose of the tour was to share this background with the students and expose them to some of the real-life challenges of forest management. Weighing the options for management in real time was important. The tour leaders asked each student what he or she would do with this seemingly healthy black ash forest. Leave it alone? Wait for the EAB to arrive? Hope that nature would find a cure? Would they preplant seedlings in the wet ground? What species might survive the water and the deer? White cedar? Balsam fir? Silver maple? Bur oak? Aspen? Red maple? White spruce? Numerous questions were posed and discussed.

I was there as an observer, but also as part of the team that had spent time on the site. We had put in regeneration plots and examined three years of natural regeneration following group selection harvesting. We were studying which species survived the small-patch harvesting and which failed. Already nature had provided some surprises. All of the conifers were doing poorly, despite their presence in the older forest. This was puzzling. Even more puzzling, however, was that some hardwoods that were not present in the stand prior to disturbance had seeded in and were doing well. Red maple, aspen, and bur oak were the newcomers. Where had they come from? Would they continue to thrive past the seedling stage?

A number of restoration topics came up for discussion that day, but the most important aspect of the conversation was the audience. We were there to make a connection with the next generation of resource professionals, a cultural connection with those who had the most to gain and the most to lose from the loss of this ash forest.

A Veteran Hunter

Long before humans were attracted to the northern forest for recreation, they came for the hunt, and their cultures depended upon their success. Tribal hunters stalked the mastodon and bison. Early European voyageurs endured arduous journeys by canoe and faced severe winters in their quest for beaver, fox, and mink. Even today, in November, a fall migration from the city brings thousands of hunters to the northern woods. Clad in orange, these modern warriors tramp through the forest in search of deer, grouse, and bear. Although not a hunter myself, I have gained respect for this cultural phenomenon. It is a ritual with prehistoric roots and some hunters come to the woods with restoration on their minds.

* * *

In my youth, when I was a vegetarian, the fall hunt bothered me. When I spotted orange ghosts walking at Esden Lake, my blood pressure rose. After all, the hunters had not been invited to my land; they were trespassing. When I asked these neighbors to depart, sometimes I was polite. But sometimes I tired of their excuses, and requested them to leave in harsh or hurried tones. Not all them respected my wishes. After all, the property had been public land in the not-too-distant past. In their minds, it was still open ground.

My response to these unwelcome guests was to up the ante. I tore down 17 wooden deer stands and burned them to ashes on the spot. Little did I realize the ramifications of this act of confrontation. Many local hunters felt threatened by this urban outsider with a different set of values. I was protecting animals in their local habitat. My neighbors were putting meat on the table in winter. The two cultural value systems were at odds, and I was clearly outnumbered and outgunned.

More than 30 years have passed since my early clashes with hunters at Esden Lake. Now, when I spot another deer stand in the woods, my eyes do not draw back in frustration or fear. That small structure represents a cultural heritage that has deep roots in preserving open space and the forest. Most members of the hunting community have a long-term investment in protecting land and its habitat. Rather than viewing hunters as opponents, I have come to see them as partners in the triangle of restoration. Over time I have observed that many hunters appreciate veteran forests as much as I do. Why would they wish to see the northern forest turned into another recreational development? This motivates them to open their wallets and invest in large tracts of forestland. Open space is preserved and forestland is protected. For 10 months of the year, these private forests lie fallow as habitat for birds and mammals, as a carbon sink, and as a source of clean water. Take a walk with a careful hunter in the woods and observe how they place their feet to avoid snags and depressions. They may hold one arm behind their back when facing prickly ash (*Zanthoxylum americanum*) or blackberries (*Rubus* spp.). They concentrate as they pause to read turkey scratching or deer scrapes.

These careful observations bring them in close contact with the forest. They learn to read and respect its signs.

* * *

Not long ago, I was in the woods with a landowner who was a hunter and woodsman. He was almost 70 years old and was a veteran from the Vietnam War. I was there to help craft a new management plan for his land, land that had been in his family for four generations. His 100-acre parcel was bordered on three sides by agricultural fields and pastureland. Tractors plied the nearby roadways and the hum of modern agriculture surrounded him. But his quarter section of woodlands and wetlands was an oasis from these activities, a haven for himself and for the wildlife he loved. The signs of animals were abundant, particularly where the western side of the property dipped to a wetland and shallow lake. Deer and beaver trails led to this secluded water source, where humans were absent.

This woodland owner had the forest in his bloodstream. For five years he had earned his living as a trapper, and he knew when the martins moved out and the possum moved in. He understood how animals traveled and what habitats they needed to survive. He was also comfortable in the swamp. Very few humans are capable hikers in wetland hummocks, where water is merely the first challenge to navigation. It takes special skills to navigate in wetlands without tripping or falling. But the bulging roots of the tamarack and spruce were no obstacles to his nimble feet. The sedges (*Cyperus* spp.) and cattails (*Typha latifolia*) parted as he slid past the suction cups of muck. By the time we reached the edge of one wetland pocket, he was a full 30 yards ahead of me, breathing lightly. He paused to let me catch up, and then he told me the following story.

The swamp was a place he knew well. After all, he had spent time in the jungle in Vietnam. During his time serving in the military, he had slept in the mud and in the warm blood of his wounded comrades. He had survived while many of his comrades were disabled or had headstones. As a survivor, this veteran knew the dangers of the jungle, and was at ease in the swamp. He wanted to share this world with his living comrades.

As we explored his woodlot, I took measurements of the trees and shrubs. But I also wanted to know what motivated him to care for the land. What would he like to see? What projects did he have in mind? He stopped leading the hike and looked me firmly in the eye. There was a short silence before he opened his mouth. He was not going to sell his forestland to a developer or turn the ground over to a nonprofit organization. What did they know about the jungle or wetlands or martins? Instead, he was planning to deed the land to his companions from the jungle in Vietnam, a local group of disabled Vietnam Veterans. He wanted them to have access to a landscape he loved, on trails he had built and maintained to handle their canes and wheelchairs. He wanted them to have a place to forget their past and their pain, a place to enjoy the moment and the wild.

I was struck by the quiet commitment in his voice. The forest was a healing place for him and he wanted to share it with his wounded colleagues. This veteran was protecting more than a piece of real estate. He was protecting a cultural connection to the land. And although he did not use the word "restoration," his intentions on behalf of his wounded comrades were designed to give something back to those who had stood beside him in arms.

Napoleon

Cultural connections come in many colors. Many years ago, I arrived late to a meeting of our local forestry cooperative. The gathering took place at a member's home in the back woods. As was the custom before the official meeting began, members sipped coffee and shared a few of their latest woodland adventures. As I slipped off my boots, one member intoned to the group, "Those guys are crazy." It took me a moment to discern to whom he was referring. I hoped it was not me.

"One year, when I was still cutting the lawn with a push mower, I had one that used to follow me around the back yard. He'd march right out there and tail behind me for a full hour or so. By the time I got done, he was plumb pooped out." The storyteller sighed and paused. "I think he was a juvenile male. You know, he had the full tail band."

Here was the clue I needed. This was a story about a ruffed grouse. The narrator noted my presence with a nod and continued. "The forester I was working with at the time told me that they liked bur oak acorns, you know, without the cap. Sure enough, he brought over some acorns and we went out and found Napoleon—that's what I nicknamed my guy, because he marched like a general (Fig. 5.1).

"Well, we found him over by his drummin' log and he went crazy over those acorns, and followed right behind us once he'd had a few. Couldn't get enough of them. Even swallowed one with the cap on. Boy, he had a tough time getting that one down!" There was another pause as the audience chuckled collectively. "Napoleon would eat right out of my hand too, but don't try to touch them; they don't like that!

"He would eat oatmeal from my hand. And he loved the warm weather, when I wore short sleeves. He'd come up to my arm and rub it a bit—wouldn't let me touch him, mind you—but it turns out he liked to pick at my skin. The storyteller rubbed his arm in remembrance of the experience and a wince of pain showed on his face.

Figure 5.1: Ruffed grouse, ©Alix d'Entremont. Used with permission.

"Really? Those beaks are *sharp*!" his voice intoned.

"One time, when he pecked me, I was a bit upset. 'Darn bird,' I said. So I just swooped him up in my arm and bear-hugged him. 'There, how do you like this? Don't be peckin' me, Napoleon. You hear me?' And when I let him go, he just backed off, kind of like a kid when you scold them. I think he really understood what I was saying, I really do."

The room was quiet, at attention. Our speaker, a former Marine, dabbed his eyes. Then he recovered his composure and changed the tone.

"And they're smart too. You can't sneak up on them, at least not on the ground. One time when I was sipping coffee on my deck, I saw Napoleon out on the gravel roadway.

'Come on, Napoleon,' I called, 'get out of the roadway.' But he wouldn't listen to me. He just stood there like a soldier at attention. And the more I spoke to him, the more rigid he stood."

There was a pause as the former Marine canvased his audience for attention, checking the ranks. "Finally, I get up. 'Okay, kiddo, what's that you see?' I asked, and I went out to the road and looked up and down. There was nothin'. Nothin'. No cars, no dogs, no nothing..."

The room was still, but our minds were working. Where was our storyteller headed?

"'Come on, Napoleon, this ain't no place for you,' I said to him. Still, he stood there, staring straight down the road. Well, I waited there with him for about a minute or so. 'What the heck?' I'm thinking.

"And just then, about an eighth of a mile down the road, two dogs came out of the woods and crossed the road. Napoleon watched them, and once they'd crossed over to the other side, he relaxed and looked at me. Well, I was dumbfounded. 'Okay, okay, now let's get off the road,' I said to him. And he heard me and scampered up the embankment back into the yard."

A spell of calm release came over the room. One member rose to refill his coffee mug. Other members relaxed back into their chairs with soft sighs. There would soon come the allotted time for the business of the day, for discussion of the newsletter and the upcoming field day. But for the moment, the cultural connection was silence and respect. Respect for our storyteller and for Napoleon, because restoration is more than a connection to the trees. It is a cultural connection with the other inhabitants that live in the forest, the third dimension of a three-sided triangle.

Cultural Collaboration

The cultural stories and experiences from the north woods remind me of the importance of cooperation in restoration. Whether on the watery old mining pits or in the silent black ash forests, human collaboration is usually a key to long-term restoration success. Sometimes the cooperators are individuals, like the veteran hunter with his passion for restoring forest trails for his war companions. Sometimes they are landowner associations or community organizations with a passion for longleaf pine or sharp-tailed grouse. Sometimes the collaborators are public managers and agents who are tasked with the job of protecting the resource. In each case, the combination of skills and cultural connections improves the odds of a successful outcome.

On private lands, most often the cooperators are outside "experts" in their field. The most common expert for small landowners is the forestry consultant. Consultants are paid to listen closely to resource issues, and then to carry out safe, practical, and beneficial projects for their landowner employers. This outside experience helps the landowners in areas where they have little or no experience or expertise. Problems are avoided and hurdles are overcome by the cooperation.

This is particularly true of restoration projects. I have been hired to restore white pine to the landscape by both individuals and landowner associations. Sometimes there are technical considerations that require significant site preparation before planting or burning. A crew may need to be hired or local officials contacted. Sometimes there are long-term technical needs, such as protection of new seedlings from mammal browse or disease. And sometimes it takes merely a knowledge of soil conditions and habitat requirements for the restoration effort to succeed. In each case, the landowner brings in outside help to reach their goals.

Often private landowners need to cooperate with public agencies in restoration projects. The soil and water conservation district technician may be called in to assess the soils for a prairie restoration or a low intensity burn. The local DNR forester or wildlife specialist may be contacted to apply for cost sharing for habitat restoration for pollinators or endangered species.

On larger private holdings, such as real estate trusts, power company lands or wood products forests, cooperation is also recognized as imperative. Many restoration projects require specialists in resource issues. The specialists may be resource managers with unique skills, such as certification protocol or water quality standards. Or they may be regulatory agents, legal teams, or elected officials who have direct influence of the outcome of the project. Often the complexity of the project determines both the extent and timeline of cooperation. Relationships are established, trust is built, and frequently the collaborators are retained for additional work.

Building relationships is a critical element of collaboration, whether for the private or public sector (Boedhihartono & Sayer 2012). As we saw with the restoration controversy in Chicago, where trust and relationships were lacking, even the best "scientific" knowledge can fall short. Science and ecology alone did not bring prairie habitat back to the public lands around Chicago. This is a sober reminder to those who serve in the academic and public sectors that they, too, need to establish relationships with those they serve. It is not enough to "educate" the public from an ivory tower and then step back. Day-to-day involvement and understanding the local needs of those most affected by the projects has to occur. Are the economic costs kept down? Will the tax base be affected by the project? What risks are involved from wildfire or flooding? Who will be around to maintain the project and answer challenges that arise after its completion? These are just a few of the considerations for public managers to consider when they design large-scale restoration projects.

Restoration of forested landscapes is never as simple as it sounds on paper or at a legislative session. There are almost always unexpected challenges, both in the field and down the line. How might future climate change affect the project? Who will carry the burden of additional or unexpected maintenance costs? How will monitoring be carried out? What are the measurable elements of success? What happens if Mother Nature throws a monkey wrench into the equation? These questions should be welcomed, not feared. And they should be asked up front so that the process is transparent to all affected parties.

* * *

A few years ago, I was called in to help coordinate a new resource management plan for the Fond Du Lac Tribal Band (FDL) in Northern Minnesota. The Band already has its own resource managers for forestry, wildlife, water quality, energy, cultural history, policy, and for the growing of wild rice. My job was to help bring each of these departments together and develop an updated integrated plan. The FDL resource history was updated, recent management projects were outlined, and ongoing challenges such as barriers to conducting restoration projects were identified. We helped develop a team approach to address complex resource issues, such as collaboration on moose habitat restoration, wild rice water quality control, and the risks of the EAB on the Band's ash forests. As soon as we had a working draft of the new plan, we took it to leadership and to members of the Band for input and review. Members were able to voice concerns at public meetings and the resource managers were able to clarify the strategies and implementation steps involved in the plan.

This process took time—almost two years—to carry out. But the importance of collaboration was clear. The managers were encouraged to work together and were put in touch with other managers and resource professionals. In addition, the members of the Band were brought in to provide their review and input. The net result was a clearer path for the future and better relationships built on trust.

To me, this process was both enlightening and a privilege. It showed me how important collaboration and cultural connections were on landscape management projects. In addition, resource managers strengthen their hand when they work together with other professionals for the benefit of all.

A Bear Story

Cultural connections to the forest are deep rooted. These roots contain human fears and hopes, and are echoed by many of our children's fables: *Bambi, The Big Bad Wolf, Goldilocks and the Three Bears*. Of all the animals in these myths, bears figure prominently. Perhaps this is because bears are strong and independent. They stand and walk and sit. They eat berries and protect their young. These mammalian attributes are both respected and feared by humans.

When friends ask about my time in the woods, they sometimes inquire whether I often see bears. This question is usually posed with some trepidation, perhaps even a sense of foreboding. It is as if I am at risk if I work in the woods. The truth is black bears are the only bears in my neck of the woods, and black bears are reclusive animals. They usually move about after dark when humans are home watching their flickering flat-screen televisions or sleeping. As a result of these nocturnal habits, along with a healthy respect for humans, face-to-face encounters with black bears are uncommon.

Not surprisingly, I have heard more bears during my days in the woods than I have seen. The sound of a bear in the woods is unmistakable. It is no wonder that the myth of Bigfoot is alive and well. The thrashing a bear makes is akin to the sound of five drunken men stumbling, complete with a few grunts for good measure. There is only one animal in the north woods, except for humans, that creates such a ruckus when it moves through the brush. And most often, just as suddenly as it is heard, it is gone. Peace returns to the forest.

But bear encounters in the forest are rare. I have seen more bears while camping with friends in public campgrounds than I have while working in the woods. For bears have a voracious appetite and are drawn to human food, barbeque smoke, and garbage cans. A good friend of mine and I still joke when recalling some of our surprise encounters with these large, limber creatures of the campground. The first time this happened we were seated at our campfire in the evening, enjoying a heated philosophical discussion, the kind that develops after a few beers. In the midst of our musings, a medium-sized black bear ambled out from behind our tent, not 20 feet away. Our initial reaction was disbelief and wonder. During a speechless pause as we pondered what to do, the bear casually lumbered away. It disappeared as rapidly as it had appeared. And we were left sitting in the same position as when it had appeared.

* * *

There is a secluded area at Esden Lake that was on the fringe of the straight-line windstorm 25 years ago. It borders a large tamarack swamp and public land to the west. After the storm I hired a logging crew to salvage the acres of my downed wood. They put in a log landing at a centralized spot where a patch of aspen had blown over. The crew used this staging area to cut, sort, and stack the salvageable wood. Large semis pulled onto the site and loaded the oak, pine, and aspen logs for nearby mills. For a few weeks that fall, it was an active operation with the constant whir of machinery and workers in motion.

When the salvage harvest was completed, the loggers moved on and the site turned silent. A large pile of unmarketable woody debris remained on the old landing. In the language of forestry, this is a "slash" pile, and these mounds of stumps and broken limbs are long-lasting reminders of many harvesting operations. The slash piles, although initially unsightly, have long-term beneficial purposes and they soon become havens for all sizes of creatures seeking shelter and protection.

When the salvage harvest at Esden Lake was complete, I seeded down that landing to clover. Within a year or two, the area transformed itself into a grassy opening surrounded by raspberry and blackberry bushes. For a number of years, I spent lazy August afternoons picking the fruit from the fine berry patch. Little did I realize that I was not alone on this landing, for a pair of large eyes was watching me from the safety of the slash pile.

* * *

September and October are special months in the northern forest. The weather is neither hot nor cold. The insect population is down. The colors shift and the sun moves lower in the sky. It is a favored season for those of us who love the outdoors.

It was afternoon as I stood on my road and listened to the laughter of my neighbor as he shared his most recent bear story. Dan was usually soft spoken, and his style was one of understatement and respect, that of an experienced hunter. For years I permitted him and his sons to hunt at Esden Lake. I would rather have had friends sitting in my trees than unknown orange ghosts. And Daniel spent countless hours patiently sitting, his bow at his side.

While Dan enjoyed the weather, his hunting skills helped to put food on his table for the winter. To my benefit, he also reduced the deer population, those aggressive ungulates who gobbled up the young white pine seedlings I carefully planted each spring. I was tired of replanting and protecting the small pines. It was far more satisfying to permit my neighbor to fill his freezer with fresh meat. Besides, although I was not a hunter myself, I appreciated the skills and patience required to have success with a bow and arrow.

Bow hunters are quiet and cautious. They have learned to be still for long periods of time. One year, in recognition of his patience and luck, Daniel was granted a state bear permit. Bear permits are hard to come by and are thus highly prized. Alone, with a few arrows, the hunter stands on the same ground only a few feet from his prey. Who weighs more? Who can run faster? This is an old test that follows serious hunters from their ancestral past.

* * *

Today Dan is the talk of the township. His face carries the easy smile of success as he recalls his hunt and his bear. And where do you suppose the bear was hanging out? For two weeks Daniel had walked past my old log landing to a spot in the woods where he awaited his prey, and for two weeks the bear had eluded him. Then, one day, he noticed something unusual. A set of bear tracks led from the tamarack swamp to the west right up to the old slash pile. The trail was fresh and the tracks were headed east, not west (Fig. 5.2).

Figure 5.2: Front paws of a brown bear.
Image by National Park Service/Kaiti Kritz.

Daniel recalls his surprise and amusement at this discovery. While he thought he had been tracking the large bear through the woods to the swamp, the bear had been resting in his den, watching the hunter pass. That bear was not stupid. Today, however, the bear is in his den no more, and his meat is in my neighbor's freezer. Daniel finally realized the strategy of that crafty fellow, and circled back to the slash pile with his bow. Of course, I too had spent hours near that log pile, oblivious to its quiet inhabitant, unaware that I was being watched. The bear never growled or harassed me. He never even gave away his secret hiding place. Why should he? He liked the berries and the halcyon surroundings, just as I did. I had provided him a home with my leftover landing, and he was taking care of the neighborhood.

Now that the bear is gone, I think it is time to place a moratorium on bear hunting at Esden Lake out of respect for their humble and peaceful ways.

CHAPTER VI

**NEW CHALLENGES TO
RESTORATION**

The Paradox of Preservation

There will always be new challenges to the forest, in the forest, and for the forest. These challenges may originate from human sources like development pressure of an expanding population, premature harvesting and high grading, or ill-conceived resource policies. They may also originate from forces beyond the capabilities of the human hand, such as windstorms, floods, or naturally occurring wildfires. There are a multitude of unknown risks to their health. On darker days, these challenges remind me of the old "whack a mole" electronic game. In this game, as soon as one mole is pushed down or eliminated from the board, a new one immediately appears in an unexpected place. The effects this game produces in its players are often impatience and frustration, and sometimes these stand as cause for abandonment of the game. Equally, there is no guarantee that restoration can be accomplished without difficulties and setbacks.

One example of the new challenges facing forestry arises from an unlikely source: the preservation movement. More than a century ago, this movement was born on the wings of American romanticism. The words of Henry David Thoreau, John Muir, and others sparked a nascent protest against the exploitation of the nineteenth century. Rivers were filling with sediment. Old forests were rapidly disappearing to the cross-cut saw. The passenger pigeon, the buffalo, and the snowy egret were being hunted toward extinction. It was time to recognize the losses and propose a new, more-protective course. In the terms of Georg Hegel, the nineteenth century German philosopher, this new course represented an "antithesis" to the old methods (Beiser 1993). Instead of subduing nature, with biblical fury, it was time to recognize its virtues and protect it.

And protect it we did. Beginning with President Roosevelt's National Forests and Parks, the movement to protect our natural heritage grew. And as it grew, it gathered a fever all its own. The new fever was to place "nature" as something not to be touched. Wilderness areas and set-asides became a rally call in Washington. Fueled by an increasingly urban population with more leisure time at its disposal, the preservationist movement has grown and flourished.

But as Hegel argued so eloquently in his dialectic analysis, any antithesis creates its own set of paradoxes and drawbacks. With the preservation movement, these drawbacks have only recently become more obvious and problematic. In the first place, preservationists still regard nature as separate from humans and human endeavors. This continued separation causes continued conflict. Instead of beneath human hands, it is now placed over our heads, almost godlike. The result is that "nature" is turned into some abstract ideal, a perfect state in need of not merely protection, but more accurately, in need of preservation. Because it is revered, it is made sacred. When nature is considered sacred, then assumptions and policies follow. "Do not touch" signs go up in the press and on public land. Human actions (including restoration) become discouraged from philosophical and political perspectives. An attempt is made to embalm nature in some culturally acceptable state, like an ancient temple. Millions of acres of national parks and wilderness areas are deemed to be out of bounds, and active management ceases there. All of these efforts are advertised in the name of a new god: Mother Nature.

But nature is not an architectural temple. It is alive and always changing. Fuel loads build. Trees die. Insect populations expand. Wildfires, earthquakes, tsunamis, and even volcanoes disrupt and degrade the landscape. And ironically the movement that set out to help protect nature mandates that nothing be done. Do not touch.

This paradox is exemplified by New England's *Old Man in the Mountain*. High above the Pemigewassit River in the White Mountains of New Hampshire rests an outcropping of granite ledges. Viewed from the right angle, these ledges once resembled the profile of an old man's face and were honored for hundreds of years. They were revered by the Abenaki culture and later that same reverence was passed to on to the new migrant settlers who inhabited the area. In recent times more than 5 million visitors were attracted to the site annually to enjoy the miraculous profile. But Mother Nature has thrown a monkey wrench into its artwork. The ledges are moving, crumbling, and splitting.

For the past 100 years, human efforts have attempted to mitigate the changes in the granite outcrop (Sanders 1997). But each year the situation becomes more challenging and currently a large chunk of the face has dropped into the valley. At what point is it best to let the old man disappear under the weight of gravity and split granite? At what point do we recognize that nature has a destructive side that we cannot control?

And there are other powerful examples that the natural landscape is changing. The ancient redwoods in northern California have been set aside in groves and parks, and yet their tops are dying. Despite our best efforts to preserve them, they are in serious decline in much of their range. Protecting them in place cannot embalm them.

The fact is, we cannot preserve nature in some idealized original condition. It is impossible to ask time to stand still, as if nature were the Parthenon, rigid and set in stone. The preservation movement, despite its good intentions and high-minded goals, has created new challenges to resource management involving how to respond to dying redwoods or how to handle more intense forest fires in areas that were set aside (fires that soon spread to inhabited areas and destroyed homes and lives). Resource managers are also confronted with challenges of wildlife habitat preservation in a time of climate changes.

Preservation is fiction for a living forest, a fiction created by the romantics. The sooner we recognize this for what it is, the sooner we can begin the next step in Hegel's model: synthesis. Hegel proposed that there are three steps to the development of philosophical standards. The first is the "thesis," or, in natural resources, the philosophy of overcoming or subduing nature. The second, as mentioned, is the "antithesis." In forestry this amounts to placing nature over our heads or out of bounds. The third stage in Hegel's philosophical analysis is synthesis. Synthesis takes elements from both the thesis and antithesis and melds them into a new practice (Beiser 1993). With resource management this new practice is called restoration.

* * *

The new approach recognizes, first and foremost, that humans are part of nature, not separate from it. We are a team, and need to work together with Mother Nature to maintain the health of our forestlands. It is a collaborative effort. It is one thing to attempt to preserve old forests, forests that will change and eventually pass away. It is another to restore and protect veteran forests to the landscape and recognize the values of mature woodlands, both culturally and ecologically. It is one thing to prohibit chainsaws or motors in wilderness areas, it is another to recognize their positive uses to reduce wildfire risks, attack invasive species, and open up hiking trails.

Because humans are part of nature, we will still face obstacles, but at least we will have discarded an impossible dream. Nature is neither kind nor perfect. She knows how to destroy as well as heal, just as we do. Restoration concentrates on the healing process. We still need to balance the costs of our labors with the ecological benefits. We still need to integrate value systems into solutions. We still need to better understand past conditions and likely outcomes of our work. But at least we will have taken off our blinders that preach that doing nothing is the best policy.

Desertification and Climate Change

Resource managers are on the front lines of climate change. The forests that we manage today are already affected by recent weather patterns, and they will be different from those of tomorrow. In fact, many biomes are already in motion, and not necessarily for the better.

One of the most striking examples of climate change is evident in the American Southwest. In fact, we might rightly call it "ground zero" in this country. The Arizona high plateau is one that is blessed with sun, but challenged by heat and a lack of rain. Most climate models show that this biome is at risk to change rapidly in the near future, and already, in the hillsides and mountains surrounding Flagstaff, the historic landscape has been affected. The first signs are evident in the small pockets of aspen that formerly regenerated on high mountain slopes and in the shadows of cool swales. These aspen clones need moisture and cool weather to put down roots since trembling aspen is a boreal species. In the past 20 years, managers in the plateau have noted a continual diminishment of these aspen stands. And they have been searching for solutions. They have even tried putting fences around a few to stave off the elk browse.

What is happening? First of all, the aspen is not regenerating on the mountain slopes. Although exact causes have not been verified, there are obvious suspects. Put simply, rainfall is down and temperatures are up.

Add to this the increasing elk population and its need for forage and you have a perfect storm working against the aspen. As a result, grasslands are replacing the high aspen forest and early signs of desertification are on display near Flagstaff.

Aspen is not the only species that is suffering in the Southwest. Ponderosa pine is also in jeopardy, but for a different reason. Ponderosa pine, like many pines, relies on fire to prepare its seedbed and reduce the competition. The seeds of Ponderosa lay dormant in the duff until conditions turn ripe for germination. However, recent wildfires are burning more intensely than in years past, and new research has shown that these hotter fires are destroying the dormant seeds and cones rather than releasing them. The result is that ponderosa pine seedlings are disappearing from the plateau landscape. While the dominant sentinel pines still silhouette the skyline, we should remember that as soon as the new seedlings are gone, a whole generation for the future may be lost.

Further north and west, there are other ominous signs of change. The tops of many ancient redwoods in northern California are dead, and their growth patterns have been reduced. Can this be related to the warmer winters or to winter fogs that are less common than they once were?

The size and intensity of summer wildfires is on the increase, and firefighting costs have blossomed as property losses mount into the billions of dollars each year. One result of these larger fires is that the U.S. Forest Service has grown strapped for funding.

In addition, mountain pine beetles (*Dendroctonus* spp.) have killed thousands of acres of conifers in Montana and Idaho, and the mills can't keep up with salvage efforts. As a result, the hillsides are filled with highly flammable fiber and the risk of intense wildfire has increased.

These are not random isolated events. They represent changes that may, unfortunately, be only the tip of the iceberg. For those who harbor doubts about climate change, I suggest taking a trip out to the western states. Observe the rust-colored mountains in Montana. Walk the hills near Reading, California or Flagstaff, Arizona. To experience this increasing mortality in our western forests is to wake up to the reality of climate change.

* * *

Not all parts of the country will suffer a loss of species as the climate changes. There will be winners as well as losers. Closer to my home, the boreal forests of the Canadian Shield are in retreat. As winters have warmed significantly over the past 40 years, boreal species, including paper birch, yellow birch, jack pine, white cedar, and tamarack, are all declining or moving northward into Canada. Meanwhile, the hardwood forests of the Mississippi valley have begun to migrate northward as well. In slow motion, many oaks, hickories, and maples have begun to move toward Lake Superior and the Boundary Waters on the Canadian border. The blue jays and the squirrels are helping them along by carrying their seeds afield.

Certain species are favored in this migration. Black cherry, Shagbark hickory, and green ash have colonized mesic sites where aspen and birch formerly held sway. What the old horticultural textbooks call Zone 4 has now become planting Zone 5. Magnolias (*Magnolia*, spp.), and honey locust, (*Gleditsia triacanthos L.*) can now survive the winters. Species diversity in my neck of the woods is on the increase.

These changes on the forest floor remind me that climate change is not a new phenomenon. For millennia, climate shifts have altered the Cascades, the shores of New England, and the deserts of the Southwest. This time it may be humans driving the change. It is time to recognize the shift and prepare to adapt for a different landscape in the future. As a result, historic conditions may not be our best reference point for restoration efforts.

There will be surprises as we adapt to new climate conditions, surprises that the computer models will miss. For example, winter temperatures have moderated significantly on the northern plains, but summers have not yet grown appreciably warmer. Early modelers did not predict that the growing season would be longer but not appreciably hotter. What does this mean? Because warm air holds more moisture than cold air, this has powerful implications for moisture-loving species. Warmer winters also affect insect survival and a host of other ecological factors, including seed stratification and pollination. Which of these factors will drive forest management in the next 100 years? Will we be prepared to handle the consequences of these changes in the future?

Although climate change presents serious challenges, it is not necessarily our enemy. There are opportunities awaiting flexible and creative management. (Nagel, Palik, and D'Amato 2017). The first opportunity is to observe how nature adapts to these changing conditions. We can collect long-term data and make inferences about growth patterns and survival. What species seem to thrive and why? Where is mortality greatest?

Our second opportunity is to assist nature with an active hand. This is the opportunity is to be proactive, rather than passive. Perhaps a certain species will do well under different moisture conditions. Perhaps a wilderness area would benefit from a prescribed burn. Perhaps our voices could sway a local park board or county commission to take action.

Climate change is one of the new variables we need to consider in forest management, but it need not be our bogeyman. By adopting more resilient strategies and prioritizing projects that address concerns over climate change, resource managers can be proactive and take the lead. This will help put us on a path of partnering with nature rather than fighting it, and is a prerequisite for success (Cubbage et al 2017, Nagel et al 2017, and Pilarski et al 1994).

Nasty New Invasives

Newly introduced insects and diseases have arrived in the North American woods. Gypsy moths have ridden over from Europe. Emerald ash borers have made the leap from mainland Asia. Both have left in their paths blight, degradation, and barely recognizable landscapes. And these are just two of the entomologists' favorite new fundraisers. For pathologists, the list is longer and more destructive: Dutch elm disease, dogwood anthracnose, sudden oak decline, butternut canker, chestnut blight, and blister rust—each from invasive fungi with elusive origins (French 1989). With each new discovery, the media chimes in, sharing information with colorful, often exaggerated descriptions, calling the newcomers to the forests "devastating" or "voracious," describing the damage they cause as "widespread," and stating that defenses against them are "nonexistent." Shades of *Chicken Little* reappear on the horizon, and yet the forests live on despite the pandemonium and widespread fears that the sky is falling.

Invasive pests are not an unheard-of threat to the landscape in North America. They are as old as human migration patterns, as old as the arrival of rats, roosters, and termites in the sixteenth and seventeenth centuries. Invasive insects are no more a surprise than lice, bedbugs, and typhoid, which also came over with European settlers.

A glance back at the nineteenth century offers an instructive lesson on invasive organisms. Chestnut blight arrived in the United States from Europe as a stealthy fungus and began its brutal march through the hardwood forests of the Appalachian Mountains. By the early twentieth century, mountain hillsides were littered with the skeletons of magnificent old American chestnuts (*Castenea dentata*). There was justified fear that the species would become extinct.

The nascent study of plant pathology was spurred on by this mysterious blight from foreign shores. Soon, the pathogen (*Cryphonectria parasitica*) was identified, then isolated and tested. Early research was accompanied by attempts to find resistance to the disease, both in the forest and with breeding techniques. Over the years, as hybridization and crossbreeding became more sophisticated, the chances for survival of the American chestnut increased. In the forest, pockets of survivors were discovered and protected, and chestnut breeders began a painstaking process of trial and error with the resistant strains. Nursery techniques grew more sophisticated and crosses and backcrosses of growing stock began to show promise. The greatest success came from hybridization efforts of the American chestnut with Asian species, particularly the Chinese chestnut (*Castenea usollissima*). Today, American horticulturists plant these resistant hybrids in parks and landscape settings across the northeast. The chestnut is on the comeback trail. Sadly, the grand old chestnuts are gone, but their passing has offered a silver lining. As a result of a powerful invasive disease, the science of tree pathology was spurred to advancement. Research efforts increased. Genetic testing improved. Restoration became possible.

These efforts represent some of the first examples of forest restoration moving from a dream to a reality. As a result, today's resource managers have better methods to protect and defend numerous tree species against blight, disease, and destruction on account of invasive species.

The American elm is another example of a tragedy turned into a restoration success story. Intensive research uncovered both chemical and biological controls for *Ceratocystis ulmi*, a European fungus that wreaked havoc on the elms, and the beetles that vectored it in the trees. At first, chemicals were applied to protect the trees. Then resistant pockets of elm were found and propagated. Nurseries now offer both hybrid and resistant elms (*Ulmus* spp.) in their inventory of landscape plants. These historic trees are on a path of slow return. Some of the old sentinels still stand, protected by newly formulated injections that kill the beetles that spread the disease. The American elm has not been eliminated from the modern American landscape. They have been bred and protected. The lengthy process of restoration has saved them.

There are other signs of the power of restoration against the spread of invasive diseases. White pine, the magnificent conifer that helped build the towns and cities of North America in the nineteenth century, is beginning to recover from European blister rust (*Cronartium ribicola*). Seventy years ago, foresters stopped planting white pine in the Eastern forest for fear of spreading the disease. Now, as a result of natural resistance and careful breeding, white pine populations are on the rise once again. The species has reinvaded old fields in New England and pasturelands in the Midwest. Silently, in the shadow of young aspen and birch it has seeded in and taken root. Within another generation, this species may be common enough to fortify the milling industry that made it famous.

Another recent example of resilience relates to the gypsy moth (*Lymantria dispar*). This foreigner was introduced to our shores by a careless Harvard researcher more than 100 years ago. He was in search of hardy silk worms to bolster New England's once-prosperous textiles and garments manufacturing industry, and brought the insects in for testing. The moths escaped and began a voracious march through oak and hardwood forests of New England and the Mid-Atlantic. As recently as 30 years ago entomologists and managers feared that the red oaks of Pennsylvania, Massachusetts, and Michigan would be completely lost to this introduced pest. But once again both humans and nature have adapted. Researchers introduced pathogens and predators in heavily infected areas (Liebhold et al). New chemical products such as *Bacillus thuringiensis* (*Bt*) were developed and tested. In addition, recent studies indicate that some hardwood forests have developed their own defenses to the insect, including genetic resistance. In short, gypsy moths have become "naturalized" in their new home. They still kill some oak trees, but the genus is no longer under a major warning. With human help, the ancient genus *Quercus* has adapted and will survive in our country.

* * *

There will always be new threats to the forest. The emerald ash borer (EAB) represents the latest example of arboreal destruction in the Midwest.

The colorful but lethal EAB is particularly challenging to find and control, mostly because of its secretive lifestyle. It prefers to breed and lay eggs high off the ground in the crotches of native ash trees (*Fraxinus* spp.), and often goes unnoticed until the woodpeckers show up. By the time the leaves and branches exhibit dieback, the EAB larvae and adults have spread into the neighborhood and often are well beyond control. But the EAB offers an opportunity for urban forest restoration. Arborists are reminded to stay away from monocultural boulevard plantings as they replant the dying ash. Foresters and researchers are encouraged to search for other species that may survive on the hydric sites that black ash (*Fraxinus nigra*) calls home. Entomologists are asked to study natural predators and parasites for this invasive insect from Asia. All of these efforts represent active steps toward restoration.

* * *

As resource managers, we also need to learn from our past mistakes. In the plant community, common buckthorn (*Rhamnus cathartica*), reed canary grass (*Phalaris arundinacea*), honeysuckle (*Lonicera* spp.), oriental bittersweet (*Celastris orbiculatus*), and garlic mustard (*Alliaria petiolata*) are all invasive species now spreading in the North American forests. Some managers advocate for eradication of these species as a means of controlling the potential consequences they may cause, but eradication has a small chance of success. Just look at the track record.

Early efforts to save white pine by eradicating gooseberries from the forest (gooseberries are a host for blister rust) did not succeed. There were too many gooseberries. Nor did it work for kudzu to spray it to death. Some seeds developed resistance. Likewise, it will not succeed to attempt to eliminate buckthorn from the forest. The birds will find it and spread it anyway. Invasive species are here to stay. As managers and landowners, we need to recognize this fact and choose our restoration strategies carefully. As arborists, we are reminded to diversify our boulevard plantings. As land managers, we are encouraged to try silvicultural trials to regenerate preferred species amidst the invasive ones. As landowners, we can learn which plants in the forest are the invasive ones and become familiar with options to reduce or control them. The threat to our forest landscape becomes an opportunity for innovation and restoration.

Land Use Legacies

We are spoiled by a temperate climate, fine soils, and low population density in the Lake States and much of America. We are spoiled by adequate moisture and a stable political system, as well as by being part of the most powerful economy on the planet. But being spoiled has its consequences.

Today, forestland in the United States is under the pressure of land-use conversion. This is an old challenge with modern ramifications. In the seventeenth century, the forests of New England were cut for wood and to accommodate grazing cattle and hogs. In the eighteenth century, forests in the south were cut to make way for tobacco and cotton. In the nineteenth century, agricultural crops replaced woodlands and prairies in the Midwest. This model of deforestation was borrowed from

European history and furrowed into our agricultural heritage. Food and wood for a growing country was a high priority, even on marginally productive lands. Millions of acres of forestland in North America fell to the axe and then the plow. By many standards this approach was fabulously successful. North America, with less than 5% of the global human population, became a supplier of food and agricultural products to a hungry world. Today, it is estimated that the farmers and food processors of North America feed approximately one-third of the world's population, which equates to more than two billion people. This is a tremendous agricultural engine, and its balance of payments has made many American farmers both respected and well to do.

An additional benefit to this success story is that Americans today spend a smaller percentage of their annual income on food than almost any culture in history. This frees up time and capital for better housing and other important amenities. It also frees us to produce and consume on a grander scale. Our bodies are taller and heavier than those of humans in the past. Our cars and trucks are larger. Our housing developments are more expansive and expensive. This last factor presents a new challenge for the nation's privately owned woodlands. With a growing population and an almost insatiable appetite for more room and more rooms, developers are eager to carve up the forest into subdivisions and second homes. In the Southeast alone, more than six million acres of forested land was converted to housing developments in the last decade of the twentieth century. This pattern is repeated from the mountains to the shining seas.

What are some of the ramifications of this expansion? With a sophisticated transportation system fueling the growth, urban centers continue to spread onto rich prairie land, calcified rangeland, or into the woods. When I descend from the air into almost any large metropolis in eastern North America today, I may witness just how far the pockets of development have spread. A window seat is the perfect spot for the reconnaissance. Forty miles out, the homes and parcel sizes start as large tracts, separated by 2 to 20 acres of privacy and trees. Here are the newer developments. As the flight reaches a 20-mile radius from the urban center, the parcels become smaller and more numerous. The suburbs begin 10 to 15 miles from the runway, split into the traditional suburban development with boulevards and scattered yard trees. Finally there is the sprawling airport complex, surrounded by warehouses and pavement. Only a few decades ago, many of these developments were wetlands, forests, or farms.

The blessings of our modern transportation system have a direct impact on the forestlands we love. As we clear breathing forests, we slowly curtail the planet's lungs and circulation system. Water quality suffers. Carbon dioxide levels rise. The canopy over our heads disappears, replaced by concrete and boulevard trees with shortened lifespans.

Land use conversion is a serious challenge, and not merely in the United States. A visit to the Amazon Basin, the largest set of lungs on the planet, confirms the global trend. Thousands of acres of rainforest have been cut and bulldozed for soybeans and cattle. Although the soils are generally poor, the growing season is long and the costs are low. Who needs another acre of dark jungle when soybeans are a worldwide source of protein?

The long-term trend of a burgeoning population and economic growth brings with it the loss of millions of acres of woodland. Government agencies can only do so much, particularly in free-market economies. Zoning ordinances and green space help, but often merely shift the development burden to another jurisdiction. Legislative tax breaks and conservation easements are options with a limited mitigating effect, given their costs and legal complexity. Even public forestlands are at risk from agricultural and mining pressure. In short, there are no easy answers or shortcut solutions to land use conversion. Perhaps some families will opt for simpler lifestyles and have fewer children. Perhaps developers will opt for cluster housing with green space. Perhaps governments will opt for tax incentives to protect forestland. Still, with population growth, the challenges of land use conversion will likely only grow more acute.

* * *

Forest restoration offers another tool to help rebalance the system. It is lower in cost than conservation easements and set-asides. It involves not merely ecology and economics, but also community action. It heals as it protects. And it does not start at ground zero. Restoration works with nature to revitalize and return health to degraded forest landscapes.

Countering the Skeptics

"America will never be destroyed from the outside,
it is only from within that we may fail."
—Abraham Lincoln

Forest health and restoration face uphill battles against land use conversion, invasive species, and climate change. However, the toughest challenge does not arise from wildfires, economics, or desertification; these threats are widely known and their battlegrounds well defined. Instead, the gravest threat arises from within, as noted by both Abraham Lincoln and Pogo: *"We have met the enemy, and he is us."*

Many landowners, resource managers, and policy makers are hesitant to approach resource management in a new light (Cubbage, O'Laughlin, and Peterson 2017). They remain skeptical of the Hegelian model of synthesis. For them, restoration is an expensive and unproven experiment. These skeptics are articulate and intelligent (Katz 1997). They have well-trained minds and offer carefully crafted arguments to support their positions and pessimism (Gobster 2000). They point to denuded forests in the tropics and the need for more food to feed a growing population. They accept greed, corruption, and political paralysis as human attributes that grease the current skids and are part of human nature. They note the ineffectiveness of national governments to address climate change even when the majority of scientists have long agreed about its deleterious effects. They emphasize the "inherently selfish nature" of human beings and assert our cultural inability to recognize the resource precipice facing civilization today. They remind us that although the cliff is straight ahead, the status quo continues and the train barrels on at high speed. What can be the solution to this headlong rush toward disaster?

Many of the skeptics' arguments are cogent and should not be ignored. There is no doubt that these are challenging times, and that cultures are slow to change. But does that mean that change is not possible and that we should throw in the towel? Perhaps this is merely an intellectual excuse not to act? "Je ne suis pas responsable" echoes from the Nuremberg trials and the history books I have read. It is a discouraging line, shirking responsibility for actions carried out in the past.

Even to argue with doomsayers often appears fruitless, except as an intellectual exercise. Why engage the dark voices of inertia if the dialogue merely provokes a defensive stance? What is gained from the debate except for an adrenaline rush and an exhausted set of lungs? Indifference is easy to justify or rationalize. Does that make it defensible?

* * *

Recently, I met up with an old friend whose dark view of the future remains unshaken. He laments the cold reality of a culture focused on immediately gratifying its desires rather than focusing on its true needs. He points out that many officials in positions of responsibility and influence still choose not to recognize, let alone address, the adverse effects of climate change and green house gases. He asks if it is possible to be optimistic about the future when so few of his contemporaries take responsibility for their wasteful ways, preferring instead to hurl themselves toward unlimited expansion. He notes that even the nation's central bankers, those who are assumed to be wise, engage in policies of endless monetary expansion. Do they assume that this is eternally possible even though it defies the laws of physics?

My friend's perspective carries weight. His arguments are difficult to ignore. But his pessimism impedes a creative response. His philosophy throws cold water on the fires of adaptation. To paraphrase his bottom line: *"there is no hope. Retreat. Prepare for the worst and stash your supplies in a safe place."*

His view is a despondent critique of his fellow men, positioned from within the bunker of defensiveness. Where has despondency gotten us in the past? Was George Washington despondent after he lost four battles in New York in 1777? To many, the Revolution was over. Was Franklin Roosevelt despondent after the attack on Pearl Harbor in 1941 when half of the Pacific fleet lay submerged? Was Rachael Carson too despondent to sit and pen *Silent Spring*?

Clearly, I am less pessimistic about the future than is my old friend. He assumes that a disastrous collapse is inevitable and his goal is to survive it. But is his bomb shelter attitude really the solution?

I do not expect to change my friend's way of thinking, nor the behavior of many others who favor the status quo. Perhaps some of their prognostications will come to pass. Perhaps the economy will falter. Perhaps the oceans will rise. But why listen too long? The skeptics have already bowed out of the game. They prefer to pass judgment from the sidelines, where they believe they are safe and secure. I believe there is another trail, a trail that harbors hope instead of despair. I prefer to engage in a dialogue and offer restoration as a viable alternative that offers to nurture our nation's forests back to health. Our forests have been cut and burned before, and have returned. The climate has warmed and cooled before, far more dramatically than we have yet experienced. Certain species have been lost, but others have adapted and survived. We ourselves are one of them. Perhaps it is our turn for a test: one more civilization to arise, flourish, decay, and be reborn. It is the rebirth that interests me. There will come a time when restoration will not only be an option, but a necessity. And then the skeptics' voices will be silent.

CHAPTER VII

A VISION OF THE FUTURE

The Benefits of Veteran Forests

"The earth has music for those who listen."

—*George Santayana*

Most of us are creatures of habit. We rise at the same hour in the morning and begin our daily rituals: a first cup of coffee, a glance at our cellphone, a shower. These rituals offer structure to our day. They provide a starting gate for the track ahead. We take a deep breath to assess the course.

Although our rituals provide a time-tested method of structure and safety, at times they may obstruct the path to progress. They may close our minds to alternatives. We may become set in old habits, focused on how we have done things in the past, unaware of opportunities for improvement.

One example of this traditional perspective is the history of even-aged management in forestry. More than 100 years ago, scientists and resource managers developed even-aged silviculture as a method to encourage regeneration of light-loving species. This new method of management also discouraged the process known as *high grading*, the scourge of harvesting only the best trees in the forest while leaving the worst behind (Smith 1986).

Foresters advanced and implemented methods of planting, releasing and growing trees of the same age class in homogenous blocks. This was a modern agricultural approach, ideal for fast-growing species such as loblolly pine (*Pinus taeda*), Douglas fir (*Psuedotsuga menziesii*), and quaking aspen. The results of this silvicultural approach were and are impressive, and the process has been widely mimicked in countries as diverse as Poland, New Zealand, Brazil, and Canada. Even-aged stand planting allows for more wood fiber to be grown more rapidly on fewer acres. Machinery is designed to cut and process the fiber efficiently. Planting stock is genetically improved to withstand diseases and insects.

One result of even-aged management is that wood and paper prices have remained relatively stable in the global economy, while feeding a growing demand for housing, furniture, and pulp. Millions of acres of young even-aged forests now flourish in the Southeast, the northwest, and the Lake States. Douglas fir and loblolly pine provide industry with the raw material it demands for global markets. Even-aged management has become a success story of significant proportions.

But as with any success story, there are unanticipated side effects. Some resource managers advocate even-aged management in almost all of their landscape decisions. The agricultural models of growth and yield fill ongoing research stations and forest managers' handbooks. Shorter rotations, where commercially viable seedlings are planted, tended, thinned, and harvested multiple times on the same site, are encouraged. Today, even-aged management dominates much of the forest industry landscape. Thousands of acres of fir, oak, aspen, and spruce forests are efficiently harvested and regenerated to a single species every year. Rotation harvesting (more commonly known as "clear-cutting") has become the new norm in much of America's Pacific Northwest and Southeast.

As these vigorous young forests sprouted across the timberlands in the United States, some people began to complain. In areas with steep slopes in the west, rates of erosion and stream sedimentation increased with even-aged management. Salmon habitat was impaired. Harvested sites took on the appearance of denuded landscapes. The increased use of powerful herbicides raised the risks of contamination to water supplies. Where mature forests once stood, young monocultural saplings took their places and pest outbreaks increased. Although many of these problems are addressed by modifications to plantation design, youth and fast growth are not always advantageous for the forest. Slower-growing species representing a mixture of age classes usually have more resistance to pathogens and more resilience in the face of climate change. Mature forests offer complementary benefits that are lacking in young forests. Certain wildlife species require them. In addition, uneven-aged forests harbor a greater variety of understory complexity and diversity (Franklin 1989). Finally, humans enjoy these forests for recreational activities including hiking, canoeing, horseback riding, and skiing, to name a few.

In the United States, we refer to many of these older forests as "old growth," and we have begun to protect them and "set them aside." But, as was pointed out earlier, set asides are not immune from change. Windstorms and fires still occur, and droughts remain a threat. Perhaps we need to adopt the British set of standards for older forests? Differentiating between classes of older forests will offer us the opportunity to actively manage many of our veteran forests.

Wildlife habitat, aesthetics, and water quality are high on the list of priorities in these management zones. Although they are not technically "set aside," these veteran forests have much to offer that is absent in younger ones (Watkins 1990). They serve ecological, cultural, and economic functions, and are frequently on managers' restoration lists. Invasive species are removed. Thinning occurs to release preferred species. Caring for the health of dominant trees is a priority to produce seed crops and fine-grained wood products.

The understory in these forests is also more complex. Shade-tolerant species are more common. Early successional species are still present, but only in natural openings where sunlight reaches the forest floor. Downed woody debris is left as a carbon sink and for wildlife habitat. Owls, woodpeckers, and bears prefer these veteran forests for their snag cavities and their cover. And the veteran forests offer a more stable hydrological cycle, with a better capacity to withstand droughts and flood. In periods of drought, they are more resilient to stress and insect infestations. In periods of flooding, they have greater capacity to retain oxygen in their root systems.

These mature forests also store more carbon than their younger counterparts. The carbon is held captive in the extensive roots, the large woody debris, and the stems. This results in lower emissions of carbon dioxide and a net carbon gain for the landscape.

Many species of high-value trees perform best as veterans. Redwood (*Sequoia sempervirens*), white pine, and white oak produce their longest-lasting material for decks, wine barrels, and tabletops. The beauty of a hardwood floor, the elegance of a fine table, the luster of an antique clock, are all derived from mature species of these veteran forests.

Finally, veteran forests are pleasing to the eye. They offer the aesthetic satisfaction of filtered light, of multiple layers of vegetation, and a greater diversity of plant life than may be found in younger, even-aged forests. They also serve as magnets for recreational activities. These veteran forests are in short supply in the United States today, and they need our help. But they do not need to be set aside with the aid of legal and bureaucratic wrangling and red tape. They merely need to be well managed with a balance of ecological, economic, and cultural goals. Many of them would benefit from restoration projects, including the removal of invasive species, thinning of fire-prone understories and protection of riparian zones.

When an architect is asked to restore a historic site, the age of the site under consideration may vary considerably, depending on what is considered "historic" for the specific geographic location. For instance, what is "historic" may be 100 years old in the Midwest, 300 years old in New England, or 500 years old in France. In Greece, Italy, or China what is considered "historic" may date back 2,000 years. Similarly, veteran white pines in Michigan may have germinated 100 years ago, white oaks in Pennsylvania 300 years ago, redwoods of California more than 500 years ago, and some bristlecone pines (*Pinus aristata*) of Arizona have been aged at more than 2,000 years. These legacy forests deserve a helping hand. They have stood the test of time. Restoration forestry recognizes their value and approaches their management with respect. Thinning a veteran forest improves its structural integrity. It is similar to bolstering up the foundation of a historic building. New trails in the forest resemble new wiring in an old kitchen; they increase enjoyment and functionality without destroying historic conditions.

Restoration forestry does not advocate more set-asides, nor does it mandate a passive approach to management. Would Western culture be better off if the Acropolis had been left to decay or if the Sistine Chapel's ceiling was permitted to fade? Would we help our heritage if Old Ironsides rotted at its wharf in Boston? It is time to step up and take an active hand with restoring our veteran forests.

A Biome Perspective

"It is not the strongest species that survives, nor the most intelligent, but the one most responsive to change."

—Attributed to Charles Darwin

Charles Darwin was correct when he posited that it is not the strongest species that survive. It is the adapters. Are we listening? Have we learned to adapt to nature's ways?

Three major ecological biomes converge in my home turf of north-central Minnesota: the Northern boreal forest, the Eastern broadleaf temperate forest, and the Great Plains prairie. This convergence puts central Minnesota in a unique position on the ecological map. Only three locations in North America are positioned at the interface of three biomes, and two of them are in the high western mountains, where few people live. The other is in my back yard and offers an opportunity to be a personal observer of climate changes.

When I drive north from the Twin Cities to the central lakes district, the entire landscape takes on a different appearance. The temperature, on average, drops 6 to 10 degrees Fahrenheit in less than 100 miles. The open fields on the Anoka Sand Plain transition to a hardwood forest. Cooler air greets me as I cross from the old prairie biome to the broadleaf forest. This forest was created as the result of glacial till soils along with a cooler climate. It appears about 80 miles north of Minneapolis where winter temperatures limit widespread crop farming.

When I continue north for another hundred miles, past Duluth, a third biome greets me. This is the sub-boreal forest and includes the Boundary Waters Canoe Area Wilderness on the Canadian border. The boreal forest is a land of birch, balsam fir, and northern pines, interspersed with wetlands and ancient glacial lakes. The hardwoods drop out of this landscape. Their plumbing is not designed for the extended sub-zero winter temperatures.

In meteorological circles it is recognized that this sub-boreal biome along the Canadian border is slowly migrating north because of increasingly moderate winter temperatures. There is extensive research and discussion about the effects of this migration, but in spite of this, there remain several important questions. What will happen to the moose population? Will hardwoods take over the landscape? Will white cedar swamps become a relic of the past?

What is less discussed is that the temperate forest in the intermediate tension zone buffers its prairie neighbor to the west and its boreal neighbor to the north. This broadleaf forest occupies a narrow ecological band from northwest Minnesota through the Driftless Area of Southwest Wisconsin and acts as a climate sponge, absorbing changes in heat and moisture. It is a tension zone for climate change. For better or worse, this tension zone is where the action is. Not accidently, it is where the tornadic winds struck my woodlands along Esden Lake in the 1990s.

Ecological, economic, and cultural influences from the west and the north meet in this tension zone. In the not distant past, the prairie Sioux battled the woodland Chippewa for the prime hunting grounds along this hardwood woodland fringe. Today, as the climate warms and the human population continues to increase, both farming and suburban development are on the increase. Many old aspen and oak woodlands have been cleared for corn or for irrigated fields and potatoes. New housing developments spread north, fueled by urban growth and the oil boom in neighboring North Dakota.

In addition, Mother Nature adds to the pressure. Straight-line winds and tornadoes have become more common as moisture from Gulf Stream winds meets the cooler boreal air. Winter temperatures have increased by an average of 9 degrees Fahrenheit in the past 30 years, pushing the boreal conifers further north. In short, the forest in the neighborhood of Esden Lake, with its pines, birches, oaks, and ash, is changing and I am fortunate enough to have a front row seat.

What will Esden Lake look like in 100 or 500 years? Will it still serve as a rearing pond for walleye, or will it have morphed into a shallow wetland of sedge and cattails? And what will the forest around it contain? Will the red and white pines that I have planted survive a warmer and perhaps drier climate? Will the paper birch and black ash disappear, to be replaced by hickory, locust, and juniper? Will invasive species like buckthorn and garlic mustard have taken over the ground layer from trillium and aster?

* * *

Roughly 55% of the presettlement forests in Minnesota are still forested. The other 45%, primarily along the tension zone, have been converted to farming or development. In addition, more than 99% of the oak savannas have been lost along the prairie interface. These savanna woodlands were open-grown forests with wide swaths of prairie grasses and forbs. They were dominated by long-lived, fire-tolerant bur oak (*Quercus macrocarpa*) and served as buffers: buffers for the buffer. Bur oak can tolerate drought, heat, and cold. It also has a tough skin. Its thick bark protects it from wildfire, insects, and disease. It produces abundant mast for a host of species, including habitat specialists like the red-headed woodpecker. The common fires from the prairie, whether set by lightning or tribal hunters, opened up the understory and reduced the population of their thin-skinned cousins, including ironwood, maple, and aspen. The bur oaks survived and came to dominate the savanna skyline. And now the bur oak savannas are almost gone.

Unfortunately, these resilient oak savannas are not well adapted to human disturbances. Their root systems are sensitive to compaction from highways, houses, and heavily grazed pastures. And they occupy good high ground, with fertile soils, making the land the perfect candidate for farming or pasture. As a result, these savannas are the most threatened ecosystem in the Midwest.

When I drive the back highways of the tension zone, I note the remnants of many degraded old oak savannas. They often occupy high knolls or ridges along the edges of farm fields. Usually they are filled with invasive understory plants like buckthorn and prickly ash. The shrub layer is dense and dark. But the overstory bur oaks are still there with their broad crowns and deeply furrowed bark. They are waiting patiently for conditions to turn in their favor.

* * *

If humans are going to adapt to future climate changes, we might do well to care for these old oak savannas rather than permit them to decay and disappear. They can handle a warmer climate. They can tolerate drought. They buffer the woodlands to the north and give many prairie species of plants and animals a place to call home. Shouldn't we begin to recognize and restore them for our sake as well as theirs?

Restoring these old oak savannas cannot be accomplished overnight. There are economic, ecological, and cultural challenges to be overcome. Oak savanna sites need to be mapped and prioritized. Funding needs to be procured from private or public sources. Prescribed burning takes a trained crew and proper timing. And the cattle need to be kept out. Where these goals can be met, where the invasive tree and shrub species are burned or felled, the savannas will return. The old bur oaks will be there to welcome and protect them.

* * *

What We Leave Behind

"The successful scientist is a poet who works like a bookkeeper...
but chastely so, taking care never to misstate facts, never
to betray nature."

—*E. O. Wilson*

Land restoration is more than planting trees and protecting wildlife habitat. It is more than seeding new wildflowers and converting old pastures. It is more than certifying our forests and purchasing products from well-managed forestlands. Land restoration is about what we leave behind.

Often the phrase, "It was my grandfather's land," spoken with an air of reverence, escapes from a landowner's lips. This simple phrase refers to more than a simple hand-me-down. It is a reference to a valuable heirloom and represents a long-term connection to the landscape and family roots. Restoration respects these roots and builds upon them. When land has been in a family for three, five, or seven generations, it carries both the heritage and the responsibility of maintaining it for the younger generation. Of course, the responsibility of generational land transfer presents both challenges and opportunities. What happens when landowners pass away or their children move on? What happens when taxes increase or a housing development goes in across the road? What happens when cold hard cash is needed to fund a college education or an upcoming retirement? Although there are no simple answers to these questions, they bear early consideration.

For example, I have transferred forestland in many fashions. I have sold it through traditional real estate agents to unknown buyers from afar. I have contracted with neighbors who wish to expand their land base. I have attached a conservation easement and offered it to friends at a reduced price. And I have placed land in a family trust, passing the decision-making tasks to my son and the next generation.

In transferring land, I have consulted with family members, attorneys, tax accountants, appraisers, friends, land trust representatives, and public land agencies. In each case, the players and circumstances of the sale were unique. With some of these forestland sales I made mistakes. Early on, when I was still in debt and paying off contracts, I hired a real estate agent to help me. He moved one parcel quickly, efficiently, and at a profit. He met with me after the sale and encouraged me to put more parcels on the market. In his eyes, ours was a profitable partnership. But a few months later I learned that he had also encouraged the new owner to split up his acquisition. Within a year, the buyer broke my old acreage into three small pieces and promptly sold them at a large mark-up. That experience left a bad taste for some realtors in my mouth.

A few years later, I explored selling land to a public agency with the idea of "protecting it forever." Surprisingly, I found that agency personnel were hard bargainers, even though the land was connected to existing public property and free from restrictions on management. These caveats worked to my disadvantage, particularly when restoration of veteran forests was not an agency priority and there were no guarantees in writing that the land would be kept in public hands. Disappointed, I ceased my discussions with the public sector.

From past experience, even conservation easements, which may offer a solution for long-term legacy, have legal and practical drawbacks. Their contracts are complex, require monitoring, and often severely limit owners' use of the land. The net result may be expensive and sometimes ends in disputes or litigation.

As a result of these experiences, I have concentrated on two more personal approaches to land transfer. They will not surprise you. One is to get to know the neighbors. The other is to keep the land in the family. If you are a landowner or family member of a landowner, do not be afraid to start a dialogue well in advance. Do not hesitate to ask probing questions that uncover opinions and self-interests. Long-term land stewardship begs consideration of the forest after you as the current owner are gone. Do you leave behind a landscape that is in better shape than you found it? Do you leave hunting memories for family and friends? Do you leave woodland that has long-term goals and is financially self-supporting? What part of you is out there, amongst the planted trees, the trails, and the "tap, tap, tap" of the woodpeckers? What part of you will live on in the future of this land?

The Strength of Renaissance

"Yesterday Nine Mile Creek was frozen fast, locked in winter, a snapshot of ice garbed in white. Today, holes have appeared in the solid crust. A gurgle of liquid reaches my ears. Mysteriously, in the night, the sounds of spring have replaced silence."

—2014 Journal entry

It has been a long winter in the north woods. The Ides of March have passed and still snow drops from the sky and builds on the ground. Where is the warm spring wind? Where is the rain? Has spring forgotten us?

Perhaps restoration of our nation's forestlands is a bit like spring this year: slow to appear and slow to take hold. But then, who had ever promised that it had a schedule and would be on time? The road to restoration is not made of yellow bricks, nor is it straight and smooth. It is more akin to a winding forest trail, littered with deadfall, rocks, and steep slopes. There are times when progress may be elusive and we need to pause. Times when it becomes apparent that cultural burdens from the past may play a strong hand against us. The risks of failure are real.

With restoration, there are no guarantees of success. Perhaps the bulldozers will arrive tomorrow. Perhaps the winds and fires will rage again next year. But these disruptions motivate for new music: a counterpoint to the silence of winter and ice. The winds will warm. The migrant birds shall return. The fire line will hold. The voices of spring sound gently in the air. The face of forest restoration is multivariant. There is no single recipe for success. Some forests will recover from fire, others from wind. Some will merely be abandoned on the floodplain. There will be losses on the front lines, casualties that motivate the development of better techniques and a greater understanding of the process.

While all is quiet in the aftermath of forest degradation, beneath the surface, hidden from view, energy moves and roots expand. One day a crack appears in the rocks, and, with it comes the possibility of renewal. This process begs for an active hand. It asks us to join in and persevere. It takes time to build a plan and send a message into the daylight. It takes time to implement the hard work of healing and rebuilding. But it is time well spent, with rewards multiplying for generations to come.

GLOSSARY

ANGIOSPERM: A plant whose ovules are enclosed in an ovary.

BIODIVERSITY: The biological diversity of a given area, whatever its size, as a measurement of the variability of species found on that site.

BIOME: A distinctive type of vegetation and climate condition of a region (e.g., a desert, a broadleaf forest, etc.).

CAMBIUM: A layer of living tissue beneath the bark that transports water and nutrients to woody plants.

CANOPY: A reference to the position of the upper strata of a forest.

CANT: A solid wood product formed from the center of a log, usually of low value.

CATKIN: A flexible scaly spike on a tree or shrub that bears unisexual flowers.

CLEAR-CUT: Harvesting operation that removes all of the merchantable woody species from a site. Also called a "regeneration harvest."

CLIMAX FOREST: An ecological term for a late successional forest that will self-maintain when not confronted by disturbance.

CLONE: A group of individuals propagated asexually from a single parent.

CONSERVATION EASEMENT: A legal instrument for reserving certain development rights on a property.

COPPICE: A forest stand arising predominantly from sprouts.

COHORT: A community of species that originated during the same time period, often in the same year.

CRUISE: A forest inventory usually concerned with merchantable wood volumes.

DEGRADATION: A reduction in the quality or value based upon a historical point of reference.

DESERTIFICATION: The ecological process whereby an area turns into a desert. It is usually formed by changes in weather patterns and moisture conditions.

DISTURBANCE: An act that destroys the tranquility of a steady state. In forestry this usually, but not always, results in degradation of the resource.

ECOLOGY: The branch of biology dealing with the relations between organisms and their environment.

EVEN-AGED: Forested area for which the overstory originated within the same time period.

FRUIT: A seed-bearing product of a plant.

GENUS (singular); **GENERA** (plural): A collection of closely related species of a plant family.

GROWING STOCK: Measurement of live merchantable wood fiber in a given area.

GYMNOSPERMS: One of 12 families of tree species that exhibit a naked seed. The conifers are the most significant family of gymnosperms.

GREEN WOOD: Raw fiber that has not been air-dried or kiln-dried.

HABITAT: The space, resources, and conditions for species to complete their lifecycle.

HIGHGRADE: Process of removing the best-quality saw-timber trees in a harvest.

HYBRID: A cross, usually between two species.

INDICATOR SPECIES: A group of species that, taken as an association, share a common ecological "niche" in the landscape.

INVASIVES: Non-native species that cause harm to an ecosystem.

INTERMEDIATE STAND TREATMENT (IST): A silvicultural treatment conducted in the middle of a stand rotation cycle.

LANDING: Designated location for harvesting operations to cut, pile, and load forest products for hauling.

MESIC: Adapted to a moderately moist habitat.

MORAINE: Glacial landform left by stagnant or receding ice fields.

NATURAL REGENERATION: Seeding without human intervention.

NATIVE PLANT COMMUNITY (NPC): An ecological classification system of plant communities based on their soil and moisture requirements.

NUT: A one-seeded fruit with a bony or leathery wall, usually encased in a husk.

PRIMARY SUCCESSION: The natural replacement of an unforested area by a forested cover type (e.g., prairie to woodland).

REGION: A large contiguous surface area often characterized by distinctive plant life.

RESILIENCE: The ability to recover from an illness or disturbance.

RESTORATION: An act of restoring; an agent that helps return health, strength, or consciousness; to bring back to original condition.

RESTORATION FORESTRY: The active management of degraded forest systems to return their health, productivity, and resilience to a predetermined goal.

RETURN ON INVESTMENT (ROI): One form of analysis used to determine economic value.

ROTATION: The full life cycle of a single crop of harvestable trees.

SCARIFY: Disturbance of duff layer to expose mineral soil.

SEQUESTRATION: To remove or set apart for storage (e.g., carbon storage).

SILVICULTURE: The art and science of tending a forested landscape.

SKIDDER: Harvesting equipment that "skids" the logs from the forest to the landing.

SLASH: Debris from harvest that is not merchantable.

SPOILS: Refuse material removed from an excavation.

STAND: Spatial delineation of a similar forest type. Usually 5 to 200 acres in size.

STOMATE: A specialized orifice on the epidermis of leaves.

SUCCESSION: The natural progression of species occupation of a site.

SUSTAINABLE: To give support or relief to; to nourish; to buoy or endure.

TAIGA: The world's largest biome. It includes the northern boreal forest and tundra.

TEMPERATE FOREST: A forest that grows in moderate latitudes and is neither tropical nor boreal.

TERRIORE: French for "territory." Historically used by viticulturists to denote soil types.

TILL: Glacial debris from an advancing or receding ice field.

UNEVEN AGE: A forest that originated over more than two different time periods.

WATERSHED: A region or land area drained by a single stream, river, or drainage network.*

XERIC: Characterized or adapted to extremely dry habitat.

From SAF's Dictionary of Forestry (2019).

BIBLIOGRAPHY

Aaseng, Norm E., John C. Almendinger, K. Rusterholz, D. Wovcha, and T. Klein. 2003. *Field Guide to the Native Plant Communities of Minnesota: The Laurentian Mixed Forest Province.* St. Paul, Minnesota: Minnesota Department of Natural Resource.

Frederick C. Beiser, ed., *The Cambridge Companion to Hegel* (Cambridge: Cambridge University Press, 1993).

Blackburn, Benjamin. 1952. *Trees and Shrubs in Eastern North America: Key to Broadleaved Species.* New York: Oxford University Press.

Agni Klintuni, Boedhihartono and Jeffrey Sayer. 2012. "Forest Landscape Restoration: Restoring What and for Whom?" in *Forest Landscape Restoration: Integrating Natural and Social Sciences,* eds. John Stanturf, David Lamb, Palle Madsen (eds.). Springer Science+Business Media. Dordrecht.

Daniel B. Botkin. *Discordant Harmonies: A New Ecology for the Twenty-first Century* (London: Oxford University Press, 1992).

Carson, Rachel. 2002. *Silent Spring.* New York: Houghton Mifflin.

Clements, Frederic. 1928. *Plant Succession and Indicators.* New York: The H. W. Wilson Company.

Cornett, Meredith and Mark White. 2014. "Forest Restoration in a Changing World: Complexity and Adaptation Examples from the Great Lakes Region of North America." In *Managing Forests as Complex Adaptive Systems,* edited by Christian Messier, Klaus J. Puettmann, and K. David Coates. New York: Routledge.

Cubbage, Frederick, Jay O'Laughlin, and M. Nils Peterson. 2016. *Natural Resource Policy.* Long Grove IL: Waveland Press.

Dobbs, David and Richard Ober. 1996. *The Northern Forest.* White River Junction, Vermont: Chelsea Green Publishing Company.

Eden, Sally. 1996. "Public Participation in Environmental Policy: Considering Scientific, Counter-scientific and Non-scientific Contributions." *Public Understanding of Science* 5: 183-204.

Egan, Dand and Evelyn A. 2001. Howell. *The Historical Ecology Handbook.* Washington, D.C.: Island Press.

Elliot, Robert. 1994. "Ecology and Ethics of Environmental Restoration." In *Philosophy and the Natural Environment,* edited by Robin Attfield and Andrew Belsey, *31-45.* Cambridge: Cambridge University Press.

Franklin, J. "Toward a New Forestry." In *American Forests*, 37-44. Nov 1989.

Franklin, Jerry F., Robert J. Mitchell, and Brian J. Palik. 2007. *Natural Disturbance and Stand Development Principles for Ecological Forestry.* Northern Research Station: USDA Gen-Tech NRS-19.

French, D. W. and Ellis B. Cowling. 1989. *Forest and Shade Tree Pathology.* St. Paul, Minnesota: University of Minnesota Press.

Galatowitsch, Susan M. 2012. *Ecological Restoration.* Oxford: Sinauer Associates.

Gobster, Paul H. and Bruce E. Hull, eds. 2000. *Restoring Nature: Perspectives from the Social Sciences and Humanities.* Washington, D.C.: Island Press.

Hardin, James W., Donald J. Leopold, and Fred M. White. 2000.
  Harlow and Harrar's Textbook of Dendrology, 9th Ed. New York: McGraw-Hill.

Hull, R. B. and D. P. Robertson. 2000. "The Language of Nature Matters." In *Restoring Nature,* Edited by Gobster and Hull, 97-118. Island Press.

Johnson, Paul S., Stephen R. Shifley, and Robert Rogers. 2002. The Ecology and Silviculture of Oaks. New York: CABI Publishing.

Katz, Eric. 1996. *Nature as Subject: Human Obligation and the Natural Community."* Lanham, Maryland: Rowman and Littlefield Publishers.

Kotar, J, J. A. Kovach, and C. T. Locey. 1988. *Field Guide to Forest Habitat Types of Northern Wisconsin.* Madison, Wisconsin: University of Wisconsin.

Lansky, Mitch. 1992. *Beyond the Beauty Strip.* Gardiner, Maine: Tilbury House.

Light, Andrew and Eric S. Higgs. 1996. "The Politics of Ecological Restoration." *Environmental Ethics* 18: 227–247.

MacCleery, Douglas. 2011. *American Forests: A History of Resiliency and Recovery.* Durham, North Carolina: Forest History Society.

Maser, Chris. 1989. *Forest Primeval: A Natural History of the Ancient Forest.* San Francisco: Sierra Club Books.

McPhee, John. 1982. *Basin and Range.* New York: Macmillan.

McKibbon, Bill. 1989. *The End of Nature.* New York: Anchor Books.

McNeely, Jeffrey and David Pitt (eds.). 1985. *The Human Dimension in Environmental Planning.* London: Croon Helm.

Michaux Andreas and T. Nuttall. *The North America Sylva. 5 vols.* Philadelphia PA: Rice and Hart. 1857–1859.

Muir, John. 1916. *A Thousand Mile Walk.* Boston: Houghton Mifflin.

Nagel, Linda M., Brian J. Palik, Michael A. Battaglia, et al. 2017. "Adaptive Silviculture for Climate Change: A National Experiment in Manager-Scientist Partnerships to Apply an Adaptation Framework." *Journal of Forestry* 115 no. 3 (May): 167–178.

Nowacki et al. 1999. *"Restoring Oak on National Forest Lands."* Washington, D.C.: USDA General Technical Report NRS-P-46.

Oliver, Chadwick D and Bruce C. Larson. 1996. *Forest Stand Dynamics: Update Edition.* New York: John Wiley and Sons.

Perlin, John. 1989. *A Forest Journey: The role of wood in the development of civilization.* Cambridge, Massachusetts: Harvard University Press.

Pielou, E. C. 1988. *Northern Evergreens.* Ithaca, New York: Cornell University Press.

Pilarski, Michael, ed. 1994. *Restoration Forestry: An International Guide to Sustainable Forestry Practices.* Durango CO: Kivaki Press.

Puettmann, Klaus J., K. D. Coates, and Christian Messier. 2009. *A Critique of Silviculture: Managing for Complexity.* Washington D.C.: Island Press.

Shrader-Frechette, Kristen S. and E. D. McCoy. 1993. *Method in Ecology: Strategies for Conservation.* New York: Cambridge University Press.

Smith, David M. 1986. The Practice of Silviculture: Applied Forest Ecology. 8th Ed. New York: John Wiley & Sons.

US Forest Service. *Suitability of North American Tree Species to the Gypsy Moth.* By Andrew M. Liebhold, Kurt W. Gottschalk, Rose-Marie Muzika, Michael E. Montgomery, Regis Young, Kathleen O'Day, and Brooks Kelley. 1995. USDA Gen. Tech Rep NE-211.

Watkins, Charles. 1990. *Britain's Ancient Woodland: Woodland Management and Conservation.* London: David & Charles.

ACKNOWLEDGMENTS

First and foremost, my gratitude extends to my editor, Jennifer Kuhn, and to the staff at the Society of American Foresters for believing in this project and guiding it through its long birthing process.

There are many other individuals who also contributed to the final product. Dan Kunde and Gary Bradford permitted me to share their personal stories despite their early reservations about "going public." Jack Goldfeather listened patiently to my early musings and remained steadfast in his support. Thomas Higgininson and Fred Clark both reviewed drafts of the manuscript and made valuable suggestions to the author. Bob Tisdale took a careful pen to the manuscript and made editing improvements beholden of an English professor. Dave Bubser and the careful peer reviewers, whose identities are still unknown, helped to keep my language clear, careful, and precise. And finally, my angelic spouse Nancy was persistent in her unwavering allegiance and her soft words of encouragement in times of doubt.

Thanks also to all others who helped make this project possible and whom my limited memory has lapsed in name.